Carrongrove
200 years of papermaking

Carrongrove
200 years of papermaking

Carrongrove Paperboard Mill
Denny
Stirlingshire FK6 5HN
Tel 01324 826462

First published 2000
Argyll Publishing
Glendaruel
Argyll PA22 3AE

British Library Cataloguing-in-Publication Data.

A catalogue record for this book is available from the British Library.

ISBN 1 902831 16 0

Origination
Cordfall Ltd, Glasgow

Printing
ColourBooks Ltd, Dublin

Illustration on title page

This etching by John R Barclay of Carrongrove Paper Mills is one of the earliest images still in existence.

The artist created a collection of drawings and etchings of industrial premises in the mid-1800s.

Cover illustrations

(Front) Foreman William Todd supervises the paper machine at Carrongrove in the 1930s

(Front flap) Kiln drying of paper at Carrongrove, 1910

(Back) A plan of Tamaree from Court of Session papers, 1831, showing the dispute between paper mill owner Weir and neighbours McRobbie and Glenny over access to Carron water
Courtesy of the National Archives of Scotland RHP 44844

Inveresk papers used in this book:

Cover	Gemini C2-S Artboard 250gsm
Endpaper	Colormaster 120gsm Pillar Box Red
Pages	Fulmar Ultra White 135gsm

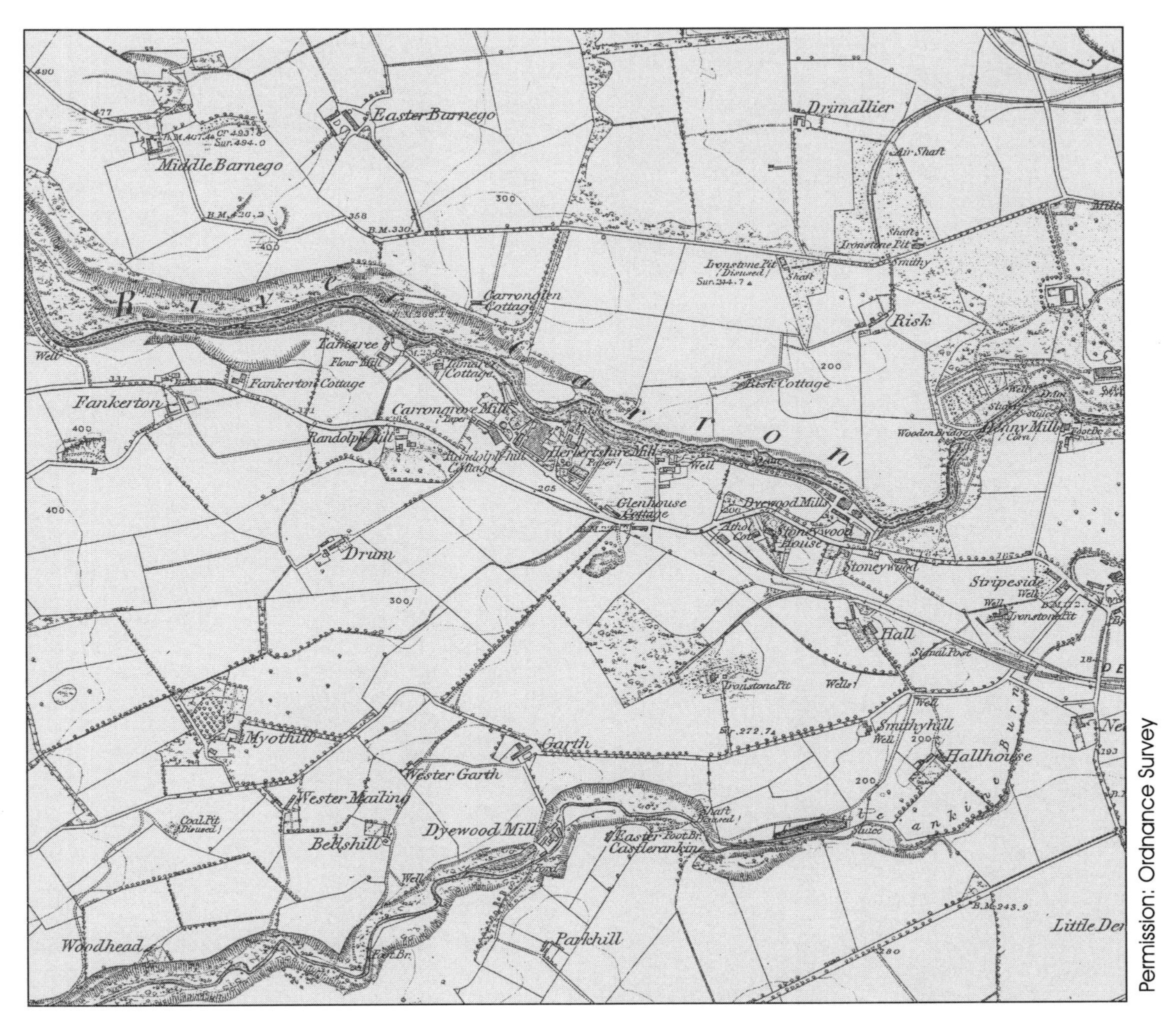

The map of 1860 showing Carrongrove Mill and surrounding farms

Permission: Ordnance Survey

(top left) Tom Marshall, Chief Engineer; William Wallace, General Manager and Director; John Herd, Foreman – pictured at Carrongrove 1930

(top right) two paper coating machines

(bottom left) a 1962 aerial view of the mill

(bottom right) no. 3 papermaking machine installed around 1900 and operational until the mid-1970s

Contents

Acknowledgements

MANY thousands of people have contributed in some way over the last two centuries to make possible Carrongrove Mill's formidable history. Carrongrove has been and continues to be a major employer in Stirlingshire and very many families have been involved in papermaking in one form or other.

In the compilation of this book, invaluable assistance was given by staff of the National Archives of Scotland, most notably by Margaret McBryde of the Publications and Education Branch, and by Linda Ramsay and Hazel Robertson in the Conservation Section. Several of the early pictorial records were also supplied by the National Archives of Scotland.

Elspeth Reid, Museums Archivist at Falkirk Council in the History Research Centre, Callendar House provided further invaluable access to precious records and some items from their pictorial collection are used with their permission in the book.

For consultations on points of Carrongrove history, including photos, thanks are due to former chief papermaker, Tom Stein; former plant technical manager, Hamish Scott; and finishing supervisor, George Burrowes.

For help with pictures, acknowledgment is due to Bob McCutcheon, The Book Shop, Stirling for kind permission to use two prints from the McCutcheon Stirling Collection. Mrs Sheena Bowden, Ian Graham and and Ivy White are also due thanks for help with picture research and kindly supplying photos. Finally, thanks to Bob Hughes, General Sales Manager Carrongrove Mill, who pulled it all together.

Foreword

I FIRST remember visiting a paper mill at about the age of seven, after my family moved to a mill house in Penicuik – my father had joined Alex Cowan & Sons Ltd in 1949. My acquaintance with mills has covered sixteen years in mainly unofficial ways and over thirty years since professionally.

The story of an individual mill is in so many ways also the history of the town or village where it is located, and from which it draws so many of its workforce. It is also a story of economic cycles, for the paper industry has always seemed prone to 'a hunger or a burst', feast or famine; and of the effect of wider influences in the world – wars, whether trade or conflict of arms, new technologies and social developments. What is happening today in Asia or North America has an effect tomorrow on the paper business. It has not needed the electronic revolution to make paper a global business.

Stefan G Kay OBE

I first peered over the fence at Carrongrove Mill in 1963 at the beginning of a cycling holiday around Scotland, at the start of the long climb up to the Carron Reservoir. Little did I know that twenty five years later, I would take up office as General Manager for a short period before becoming General Manager of Inveresk.

Some paper mills seem to have the gift of survival, of changing what they do just in time to avert certain closure. Carrongrove certainly has exhibited this quality over the years, being early into the manufacture of high quality coated papers and, when scale of production overwhelmed smaller mills in the 1970s – including Alex Cowan & Sons Ltd – shifting over to heavier weight paperboard manufacture, but still of the highest quality, and being kept up to date as technology and fashions change.

Not least is this a story over more than two hundred years, of people, for a paper mill is a dull, dank and gloomy place if the machinery is shut down and there are no people making it work.

The vision of successive managers to take a chance on the way ahead, and to invest in new equipment and new markets, has been paramount. It is always easier, and less trouble in the short term, to rely on the old ways. Carrongrove – and all mills in the United Kingdom – would simply be a memory if new ways had not been adopted.

Robert Hughes, General Sales Manager at Carrongrove Mill, encouraged by both David Ferguson, General Manager, and his predecessor Duncan McLean, has long taken an interest in the mill's history. This has been greatly helped by Margaret McBryde at the National Archives of Scotland, and Elspeth Reid of Falkirk Council's History Research Centre, who have uncovered fascinating tales and trails across a time span which runs from the eighteenth to the twenty first centuries.

I believe that not only will this book prove as absorbing to the casual reader as it does to the historian or to the paper maker, but it will be a source of valuable historical data for many years to come. I hope that you will enjoy it.

Stefan G Kay OBE
Managing Director
Inveresk plc

Der Papyrer.

Ich brauch Hadern zu meiner Mül
Dran treibt mirs Rad deß wassers viel/
Daß mir die zschnitn Hadern nelt/
Das zeug wirt in wasser einquelt/
Drauß mach ich Pogn auff dẽ filtz bring/
Durch preß das wasser darauß zwing.
Denn henck ichs auff/ laß drucken wern/
Schneweiß vnd glatt/ so hat mans gern.

F ij   Der

This print showing early papermaking is from a book of trades *Eigentliche Beschreibung alle Stände auf Erden . . .* published in Frankfurt am Main in 1568. Early papermaking in Scotland was much influenced by French and German tradesmen. The text translates –

Rags are brought to my mill
Where much water turns the wheel;
They are cut and torn and shredded;
To the pulp water is added;
Then the sheets 'twixt felts must lie;
While I wring them in my press.
Lastly hang them up to dry;
Snow white in glossy loveliness.

Carrongrove

200 years of papermaking

TWO centuries ago, papermaking on the River Carron near the village of Denny in Stirlingshire was virtually a one-man operation started by the owners of the Herbertshire Estate. The one vat paper mill, known as the Herbertshire Mill, was built at a cost of £500 and paper was produced by leaving rags, which had been bleached and washed, to ferment and decompose.

The resulting mulch was then pounded with iron rods, propelled by water power, to produce a pulp. This pulp was then left for several weeks to mellow in boxes.

This original method of papermaking was very lengthy and produced very little paper at the end of the process. But due to increased demand by 1818 there were two mills operating on the River Carron. The second mill took the name of Carron Grove and employed 25 people making a coarse paper board from old tarred rope. The Herbertshire Mill meanwhile employed 72 people and 4 horses to make a fine writing paper, printing and envelope stationery.

In the following years the industrial revolution, as well as creating a demand for paper and board, was producing great social changes. Mill activity saw the growth of Denny – by 1834, the village had expanded to a population of 4,300. As technology improved and new paper machines were installed, so the output of the mills increased. By the 1870s production was around 20 tons per week and around 200 people were employed.

International connections were established very early. Robert Weir, an Edinburgh businessman and one of the mill's first owners, established a papermaking and stationery empire across Scotland and then across the Atlantic.

The Historical Background

PAPERMAKING was well established in England and Europe during the fourteenth and fifteenth centuries but was not so in Scotland until the eighteenth and nineteenth centuries. In the meantime, Scotland imported paper from England, Germany, France and Holland, the coarse grades of brown for wrapping and white for recording legal, financial and commercial transactions.

'King James VI and the Scottish Parliament offered inducements to foreign craftsmen to train local workmen in their craft'

King James VI and the Scottish Parliament, recognising the need to establish new industries to compete with their more successful neighbours, offered inducements to foreign craftsmen to settle in Scotland to manufacture and train local workmen in their craft. The necessary skills and expertise in the art of papermaking were thus introduced to Scotland during the seventeenth century by experienced papermakers from Germany and France.

Twelve papermills were established in Scotland before 1700, seven in the Edinburgh area, two in Glasgow and the remaining three in Aberdeen, East Lothian and Berwickshire, producing around 150-200 reams of paper per week. Faced with irregular supplies of water and rags and a shortage of skilled craftsmen, these early attempts at setting up papermills in Scotland were however short-lived.

The earliest known paper mill in Scotland can be traced back to the late sixteenth century to Mungo Russell and his son Gideon who, around 1590, owned Dalry Mill on the Water of Leith in Edinburgh.

In 1588, James VI granted liberty to Peter Groot Haere from Germany *to set up this art of making paper of all sorts within this realm*. Two years later, Peter Groot Haere and Michael Keysar, *almanis* [German] *paper makeris* to the King were granted a monopoly for nineteen years.

Keysar and another German, John Seillar, entered into a contract with Mungo and Gideon Russell in 1594 for the lease of Dalry mill for eleven years for the production of hand made paper. Keysar and Seillar agreed to instruct apprentices chosen by Gideon who in turn extended the mill accommodation and installed a drying loft. Keysar and Seillar were granted

> *. . . all and haill the West Mylne of Dalry lyand within the Schirefdome of Edinburgh quhilk is presentlie a paper mylne with the dam wattergang frie ische exit and entre and all uther commmoditeis eismentis and pertinentis pertening mylnes. . .*

while Gideon Russell was to

> *. . . make ane sufficient loft of the haill lenth of the new hous . . . and sall mak twa windois in the seids loft and put cordis thairin for hinging and drying of paper therupoun. . . to the effect the saidis Michaell and John may haif and occupie the said loft and half of the said laich hous with the said mylne.*
>
> **[NAS: Register of Deeds: RD1/50 f342]**

'mak twa windois in the seids loft and put cordis thairin for hinging and drying of paper therupoun'

It is thought the original Dalry Mill failed around 1605 and nothing is heard of it again until 1673 when it was leased by six Edinburgh merchant burgesses who introduced French craftsmen to manufacture paper.

Canonmills was another early mill referred to in Edinburgh Town Council minutes of 6 July 1659. The magistrates of Edinburgh appointed the Dean of Guild, Treasurer and other officials *to visite John Patersones paper milne at the Cannounmilnes and the dame theirof and to report.* [Extracts from the Records of the Burgh of Edinburgh, 1655-1665, (Edinburgh, 1940), p154]

The mill was still in production in 1681 under Peter Breusch, a German engineer, but he complained to the Privy Council after attacks on his mill, property and staff by disenchanted locals. The mill ceased production and he moved his operation to Restalrig.

Years later he was appointed printer to the Royal Household by James VII.

Government attempts to promote new industry continued in the seventeenth century. In 1661, Parliament passed an act setting up a Council of Trade for this purpose. The import of necessary raw materials was to be free of duty as was the export of manufactured goods. To encourage the influx of foreign craftsmen, they were allowed the same privileges as native inhabitants.

On 22 January 1675, Archibald Home, merchant and then owner of the paper works at Dalry, petitioned the Exchequer Commissioners to support the development of his business.

> *. . . the petitioner haveing upon his great expence brought home frenchmen and erected ane manufactorie for paper and playing cards and have attained to a considerable perfectione in both as parcells of them readie to be produced befor your lordships will instruct.*
>
> *. . . May it therefor please your lordships upoun consideratione of the premissis (and that ther are sevinteen scotsmen and boyes bred up and instructed in these airts be the french which may tend much to the publict benefit of the kingdome) to declaire the forsaid wark ane manufactorie and to grant them all powers priviliedges and immunities belongeing to ane manufactorie within this kingdome.*

In response, the Lords Commissioners of his Majesty's Treasury and Exchequer declared

> *. . . all such materials quhatsoever usefull for that manufactorie imported [or to] be imported to be frie of the dueties of custom and excise and ordane ther act to pass therupon.*
>
> **[NAS: Warrants of the Exchequer Register: E8/26]**

Nicholas Dupine, a French refugee, established the Scots White Paper Manufactory in 1695. He floated the company with capital of

£5000 sterling and the support of the Privy Council which granted him the privilege of carrying the watermark of the Scottish coat of arms. Following this a contract, dated 16 August 1695, recorded 4 April 1698, was agreed between Nicholas Dupine, Denis Manes and others to establish two paper mills at Yester in East Lothian and on the Braid Burn in Edinburgh and to train ten apprentices. Dupine and Manes undertook

> *. . . to oversee the building of tuo papermylnes for makeing of whyte wryting and printing paper for the use of the said Company and the buying and furnishing of all materialls necessarie for the saids mylnes which mylnes are to be built . . . at Yester and the other near Edinburgh whair the air and water shall be found most agreeable for the makeing of good and sufficient whyte and printing paper.*

and to train

> *. . . tuo overseirs whose work is att the fatt [vat] tuo cocheers tuo leveers and tuo that lookes after the rags the morters and beating stuff commonly called governours and tuo masters of salls which size and finish the paper.*
>
> **[NAS: Register of Deeds: RD3/89 ff117-118]**

Crippled by debt, the Scots White Paper Company was sold to new owners around 1703.

The Act of Union of 1707 had a profound effect on the development of the paper industry and economy of Scotland as a whole. With increased political stability, once unrest had died down after the Jacobite rebellions, new mills were set up taking advantage of the English market, family businesses started to thrive, new chemical and mechanical processes were introduced and the paper industry prepared itself to take on a position of importance in the nineteenth century.

Early Papermaking in Scotland

THE basic requirements necessary to set up a papermill were a plentiful and reliable source of water to power the mill and wash and treat the fibres, a good supply of rags from well-populated areas and outlets to sell the finished product. The majority of mills throughout Scotland were therefore established in or near river valleys, where falls in level could be exploited by dams and waterwheels, seaports for access to the import and export trade, and towns for supplies of raw materials and ready markets.

'the linen rags . . . are fermented till they sweat and rot'

A typical one-vat mill, on average, could produce two to three hundredweights of paper per week. Water from the nearby river drove the water wheel that powered the main shaft fitted with a series of cams which lifted and dropped the stampers for breaking down the rotten rags in the pulping troughs. The vatman dipped wire moulds into the vat of pulp shaking out the excess water to form matted layers of fibres to make the sheet of paper, a delicate operation requiring great skill and practice. The coucher and levermen transferred the sheets of newly formed paper from the hand mould to a felt blanket where surplus water could be pressed out in a screw lever press. The sheets of paper were then hung up to dry in the ceiling area suspended on horse or cow hairs to be sized and finished, cut and parcelled for sale.

A contemporary eighteenth century account of the art of papermaking published in the *Universal Magazine*, in June 1752 describes the standard process in great detail:

> The linen rags . . . being carried to the mill, are first sorted, then washed very clean in puncheons whose sides are grated with strong wires and the bottoms bored full of holes. After this they are fermented, that is, laid in square heaps, close covered with sacking, till they sweat and rot, which is commonly done in four or five days.

When fully fermented, they are twisted into handfuls, cut small, and thrown into oval mortars made of well-seasoned oak, about half a yard deep, with an iron plate at the bottom an inch thick, 8 inches broad and 30 long. In the middle is a washing-block with five holes in it, and a piece of hair sieve fastened on the inside, so that nothing can pass out except dirty water. These mortars are continually supplied with water by little troughs from a cistern filled by buckets fastened to the floats of the great wheel, which raises the wooden hammers for pounding the rags in the mortars. When the rags are beaten to a certain degree, called the first stuff, the pulp is removed into presses where it is left to mellow about a week; then it is put into a clean mortar, pounded afresh, and then removed into presses or boxes as before, in which state it is called the second stuff. The mass being beat a third time till it resembles flour and water without lumps, it is thereby fitted for the pit mortar, where it is perfectly dissolved, and is then carried to the vat to be formed into paper.

'the mass being beat a third time till it resembles flour and water without lumps'

. . . When the stuff is sufficiently prepared . . . , it is carried to the vat, and mixed with a proper quantity of water, which they call priming the vat. The vat is rightly primed when the liquor has such a proportion of the pulp so that the mould dipped into it will just take up enough to make a sheet of paper of the thickness required. The mould is a square sieve above one inch deep, having a brass wire bottom, resting on sticks to prevent its bagging, and keep it exactly horizontal. This mould the maker dips into the liquor, and gives it a shake as he takes it out to clear the water from the pulp. He then slides it along a groove to the coucher, who turns the sheet, lays it on a felt, and returns the mould to the maker, who by this time has prepared a second sheet in another mould, and thus they proceed, laying alternately a sheet and a felt till they have made six quires of paper, which is called a post. This quantity is then put under a press, and by the strength of five or six men all the water is pressed from it, after which it is separated from the felts, laid regularly one sheet upon another, and having undergone a second pressing, is hung up to dry.

When sufficiently dried it is taken off the lines, rubbed smooth with the hands, and laid by till sized, which is the next operation. For this they chuse a fine temperate day, and having boiled a proper quantity of clean parchment or vellum shavings until they come to a size, they prepare a fine cloth, on which they strew a due proportion of white vitriol and rock alum finely powdered, and strain the size through it into a large tub, wherein they dip as much paper at once as they can conveniently hold, and with a quick motion give every sheet its share of the size, which must be as hot as the hand can well bear it. After this it is pressed, hung up sheet by sheet to dry, and then, being sorted is told into quires, which are folded, pressed very hard and so tied up in reams or bundles for sale.

[*The World's Paper Trade Review*, 18 April 1913, p3-5]

'all the water is pressed from it, after which it is separated from the felts . . . and hung up to dry'

Only vegetable fibres can be used for the manufacture of paper so rags from woollen garments were of no use. Demand for cotton and linen rags was therefore high. Local collectors or dealers, usually women, made house to house collections to gather rags and sold them to the mills. The bulk of supplies, however, came from organised schemes which supplied sorted rags and commanded higher prices.

In May 1754 a Select Society for the Encouragement of Arts, Sciences, Manufactures and Agriculture was founded in Edinburgh. Part of its remit was to award prizes for inventions and improvements in connection with trade and manufacture. One of the first areas considered by the society was the manufacture of paper.

To help promote the industry, the society offered prizes ranging from of £1 to 10s for parcels of good quality sorted rags gathered within a time limit. In 1756, the scheme offered a prize of two guineas for superfine rags of muslin, cambric and linen worth 5s a stone or more. In 1757, Janet Mitchell of Tranent earned a guinea prize for collecting 606 stones of white rags for papermaking. A guinea prize was also offered to anyone from a private family collecting the largest quantity of rags. The prize went to Miss Betsy Gibson, daughter of an Edinburgh merchant. [*The World's Paper Trade Review*, 24 January 1913, p4]

Such competitions and incentives ran throughout the 1750s and 1760s and gradually public institutions and large private families sent their rags to the mills, but by 1764 the novelty had worn off and such schemes came to an end. Despite these efforts, local supplies of rags could not meet the demands of the trade. This led to the increased import of rags from Europe. During the latter half of the eighteenth century, thousands of tons of rags arrived in Leith from Germany and Italy to support the growth of the paper industry.

'only vegetable fibres can be used for the manufacture of paper so rags from woollens were of no use'

Falkirk Museums P13126

'a plentiful and reliable source of water to power the mill and wash and treat the fibres'

The plan of 1801 involved building a cut five feet wide and two and a half feet deep to deliver water with a height or fall of at least ten feet above the surface level of Herbertshire Mill. (See back cover map)

The above photo, thought to be from the 1890s, shows the mill lade – the means of taking water from the river to better control flow and supply. The bridge across the mill lade was removed when the lade was filled in at this part

Papermaking on the Carron

DENNY began life as a rural community with a population, recorded in 1790, of 1,400. By 1841, according to the *New Statistical Account of Scotland*, this village was fast becoming a busy industrial town manufacturing cloth, dye stuffs, millboard and paper. The population had grown to 4,300 and a total of eleven mills were operating on the banks of the River Carron within a mile and a half of the town itself. Of these mills, Herbertshire Mill is described as *'the oldest establishment in the parish for the manufacturing of paper'.* [1]

Herbertshire Mill was a one vat papermill built in 1788 by William Morehead, the owner of Herbertshire Estate, at a cost of £500. He leased his mill to a series of tenants whose rapid turnover of occupancy between 1789 and 1801 is a clear indication of their struggle to make a success of the business.

His first tenant was James Liddell, followed in November 1790 by Daniel Macdonald to whom Morehead agreed to supply *a set of felts and frames, a wynch and scales and weights.* [2] In 1791 Edward and Richard Collins took on the expense of building a workhouse, dryhouse, dwelling house and offices and received an advance of £100 from Morehead for more machinery. Despite these improvements, they only managed to produce six reams of paper per day due in part to an irregular supply of water from the Carron which was seldom enough to drive the engines summer and winter. [3] They soon fell heavily into debt over rent arrears and in 1795 the business passed into the hands of Francis Strachan and Gilbert Laing. Strachan and Laing inherited the Collins' problems and finding the mill somewhat run down were only able to increase production by two reams of paper per day. Being no more successful than their predecessors, the mill changed hands again in 1797 to Adam Grieve. Some three years later, a lease for the mill was advertised in the *Edinburgh Evening Courant* on 19 July 1800:

The supply of water and the maintenance of the mill lade, as these early pictures show, was a constant factor in the production of paper from the earliest years

. . . The Mill, which contains two engines, with very complete and substantial machinery, utensils etc. and the Dwelling Houses, Offices and Work-Houses etc. were all erected and complete within these few years, in a substantial manner at the expense on the whole of upwards of L.2000 Sterling and of course are at present in the best order and repair, so that there is little occasion for immediate outlay, and a capital to carry on the manufactory is only required. The situation is known to be advantageous for the purchase of rags etc., the profits on the manufacture are very considerable, the demands for produce great, with quick returns, and the rent expected is moderate.

The Dwelling House is elegant and commodious, in a situation highly romantic and picturesque, and within a few hundred yards of the great roads to Stirling, Glasgow and Edinburgh.

. . . Mr Grieve, the present possessor, will show the works etc. and for further particulars application may be made to William Morehead, Esq., the proprietor, at Herbertshire.

Charles Laing became the next tenant of the mill in 1801 when William Morehead supported the addition of a new water-wheel and in 1806, a second vat.

Morehead recognised the need to improve the water supply from the River Carron so in 1801, along with two other local

businessmen, Archibald Napier of Randolphill and John Reid of Bonnymill, agreed to construct a mill lade at Tamaree Linn provided it did not interfere with his existing paper mill. Their agreement, recorded in the Register of Deeds on 10 September 1802, involved building a cut five feet wide and two and a half feet deep to deliver water with a height or fall of at least ten feet above the surface level of Herbertshire Mill

> *. . . in such a direction as shall be found to be most suitable for all the said parties to the present mill dam of the paper mill at Stonywood belonging to the said William Morehead. . . and that the said parties shall have full power and liberty to erect such mills as they shall think proper upon the sides of the said canal each of them within his own property.*
>
> **[NAS: Register of Deeds: RD2/286 ff 54-58]**

Each was to pay for his own section of the canal and be responsible for its maintenance and repair. In effect, however, William Morehead passed his share of the expense of building the lade onto his mill tenant and paid John Reid ten shillings annually to undertake the upkeep of his section of the canal. The terms of the agreement further stipulated that Napier and Reid

> *. . . shall not suffer or allow any ashes rubbish or other nuisances to be thrown into the said canal which may be hurtful to the*

The first of an 11 page claim at court by Robert Weir against Gavin Glenny – the dispute was about access to water and alleged interference with the river upstream

OCTOBER . 1830.

ADVOCATION
ROBERT WEIR,
AGAINST
GAVIN GLENNY, and OTHERS.

MACLEAN & GIFFEN, W. S. Agents.

LETTERS OF ADVOCATION

ROBERT WEIR, Paper-Manufacturer at Carron Mills;

AGAINST

GAVIN GLENNY, Paper-Manufacturer at Carron Grove, and OTHERS.

WILLIAM, &c. WHEREAS it is humbly meant and shewn to Us by Our lovite ROBERT WEIR, paper-manufacturer at Carron Mills, in the parish of Denny, and merchant in Glasgow, complainer; THAT an action was some time ago commenced before the Sheriff-depute of Stirlingshire, and his substitute, at the complainer's instance, against Gavin Glenny, paper-manufacturer at Carron Grove, James Macrobbie, paper-manufacturer at Bridge of Allan, and Robert Macrobbie, residing there, by a petition, of which the following is a copy: 'Unto the Honourable the Sheriff-' depute of Stirlingshire, or his substitute, the Petition of Robert ' Weir, paper-manufacturer at Carron Mills, in the parish of ' Denny, and merchant in Glasgow; Humbly sheweth, That the ' petitioner is tenant, under William Morehead, Esq. of Herbert-' shire, proprietor, in the paper-mill at Stoneywood, by virtue of

Neill & Co. Printers. A

'Access to the water was to present problems for the mill owners resulting in lengthy court proceedings in the early 1830s'

washing of paper at Stonywood. . . nor. . . be at liberty to interrupt the course of the water in the said canal so as to stop or injure the operations in the said paper mill.

[NAS: Register of Deeds: RD2/286 ff 54-58]

Access to the water and maintenance of the canal were to present problems for subsequent mill owners resulting in lengthy court proceedings in the early 1830s.

In 1817, William Morehead granted a nineteen year lease to John Andrew, papermaker and Robert Weir and Gilbert Kennedy, stationers in Glasgow, for the manufacture of paper at Herbertshire Mill. John Andrew gave up his share of the lease in 1820 in favour of Weir and Kennedy who bound themselves to maintain

Courtesy: Tom Stein

Either not enough or too much!

The River Carron was known to rise 14 feet in 20 minutes during storms resulting in flooding of the mill.

It was not until 1920 that the course of the river was moved away from the machine house wall which had been built to curve along its course.

The spoil from the digging of the altered course can be seen on the higher ground to the right in this 1920 photo

> *. . . All and Whole the Paper Mill at Stoneywood now called the Herbertshire Paper Miln and whole Machinery and Utensils therein. . . with the cottage houses pertaining thereto together with the dwelling house and offices and workhouse and dryhouse over it. . . the whole valuation thereof amounting to Four Hundred & Ninety Eight Pounds & Sixpence Sterling.*
>
> **[NAS: Stirling Sheriff Court records: Register of Deeds: SC67/49/64 ff65-69]**

The lease, in addition to recording the extent of the mill at this time also contains a detailed inventory of the machinery in use. This inventory reflects the gradual development of Herbertshire Mill, during almost thirty years of William Morehead's ownership, from a one vat mill to one possessing two vats with hogs and wheels, two stuff chests, seven presses, one pair of couch planks and a dusting machine.

Robert Weir bought Herbertshire Mill from Morehead in 1824 and over the next thirty years, as one of his local business interests, saw it expand in the manufacture of writing and printing papers and envelopes.

An aerial view of the Carrongrove mill looking west in the 1960s.

This clearly shows the strategic location on the course of the River Carron.

The houses to the west of the plant were built to house mill workers on the site of a farm at Fankerton

The second paper mill to set up operations on the River Carron was named Carron Grove Mill sited virtually next door and upstream from Herbertshire on the canal lade. Few records of Carron Grove Mill survive to confirm the exact date of its establishment but there is reference to a Carron watermark found on white wove foolscap printing paper ascribed to a Thomas Burns of Carron Grove Mill in 1819.[4]

A list of *Paper Mills in Scotland* published in 1832 records the names of mills and their owners, the mill excise number, the method of manufacture (by hand or machine), and the types of paper produced.[5,6] It identifies several paper mills in the Denny area including Stoneywood Mill (Herbertshire), No 39, owned by Robert Weir producing paper by hand in two vats, and Carron Grove Mill,

No 41, owned by Gavin and later, Michael Glennie, operating one vat in the manufacture of writing, cartridge and coloured paper.

Identification numbers were allocated to mills for the purpose of levying excise duty on paper. Each mill had to carry its number on their building as well as print it on the firm's letter heading to assure customers that duty had been paid and products were safe to use. Excise officers had the right to enter mills at all hours to examine stocks and check books. These numbers do not refer to the order of origin of a mill. When a mill ceased production and closed down, its number was transferred to another which has resulted in two mills receiving the same number. This makes tracing their history more difficult.

Nineteenth Century Developments in Papermaking

CHANGES in the paper industry were inevitable but gradual. Prior to 1780, a large proportion of the population was illiterate so most seventeenth and eighteenth century mill production concentrated on wrappings rather than white paper. Between 1798 and 1861, Scotland's population doubled in size. At the same time, the country experienced an upsurge in new literary and scientific ideas, which were reflected in the growth of the newspaper and publishing trades, especially in Edinburgh and Glasgow. A total of 95 newspapers were started up between 1815 and 1860. A similar growth in the textile industry required more pressing and wrapping paper for bales of cloth and white paper for business records, correspondence and legal papers.

'by 1800, forty-two paper mills were in operation in Scotland'

By 1800, forty-two paper mills were in operation in Scotland. Over the next thirty years, with increased demand for paper both at home and abroad, notable towns with paper mills were Aberdeen, Edinburgh, Glasgow, Perth and Stirling, the lesser including Airdrie, Greenock, Linlithgow, Mid Calder and Penicuik.

Paper making machines were first introduced in the early nineteenth century at a time of great industrial change and development. Few purpose-built mills were set up to accommodate new paper machines with the exception of Townhead in Kilsyth and Overton in Glasgow built around 1826-27 and Chirnside Bridge in 1842. Landlords, wholesale stationers, rag merchants and paper-makers, who financed the establishment and equipping of paper mills, did not have to relocate to install machines but simply adapted their mills to cope with the increased output from the new machines. They already had water to drive the engines and process the paper, raw materials, transport, markets on hand and an established workforce to draw on. They needed more staff to sort and prepare larger quantities of rags and operate the machines and more building space for storage, equipment and drying facilities. Both methods of paper manufacture ran concurrently for the best part of the century, the last handmade paper mill in Scotland, Milholm, near Cathcart continuing until 1873.

By 1852, Alexander Cowan and Sons was one of the largest papermaking establishments in Scotland, running Low Mill, Bank Mill and Valleyfield Mill at Penicuik with a total of twenty-one beating engines. By 1861, they were producing 1,600 tons of paper per year and some five years later, 2,093 tons.

The advantage of the paper machine was flow-line production which allowed the manufacture of greater quantities of paper of a higher and more consistent quality at lower cost. They soon became an economic necessity. From 1845 onwards, machine mills predominated in Scotland. Machines were installed in mills at Dalmore, 1843; Levenbank, 1848; Inverurie, 1858 and Avon, 1860. Nine new mills opened around the Glasgow area alone at Govan, 1832; Westfield Mill, Bathgate, c1834; Woodside, 1837; Kelvindale, 1845; Caldercruix in Airdrie, 1848; Port Dundas, 1850; Bowling, 1852; Clyde, 1856 and Govanhaugh, c1860. As a result of this rapid expansion in the Glasgow area, output trebled in the years between 1833 and 1849. By 1890, the industry peaked with 69 mills in operation.

Customs and Excise

Paper was seen as a source of revenue by government from almost the beginning of its use and manufacture in Scotland. The Paper Duty Act, 1712, imposed tax at a certain amount per ream for eleven specified types of paper. Over the next 150 years, 26 other acts were passed altering, extending and changing the amount and method of excising paper. Between 1781 and 1793, government assessed the tax by the physical description of all papers on the market, some 73 specified types. By 1786, although Scottish output of paper had not increased markedly, the level of taxation had risen by 40 per cent.

The system of assessment was so cumbersome by this time that in 1794, the government divided all types of paper into five classes and decided in future to assess by weight. Writing, printing and drawing papers were taxed at $2\ ^{1}/_{2}$d per lb. High-class brown papers and coloured sorts at 1d per lb; common wrappings at $^{1}/_{2}$d per lb; glazed paper was taxed at 6s per cwt and finally, mill and paste boards at 10s 6d per cwt. Although the duties themselves were simplified, provision for wrapping, inspecting and stamping each bundle of paper by Excise Officers became more rigorous.

Despite these heavy duties, which were doubled in 1801, the Scottish paper trade continued to grow. In March 1801, production of 1,145 tons of paper and 40 tons of pasteboard raised £19,000 in duty. A reduction in paper duty by half in 1836 and the introduction of steam as a motive power led to a further increase in output. Paper duty was finally abolished in 1861.

The Development of the Paper Machine

Major improvements took the form of machinery invented to replace time-consuming manual processes with faster more efficient methods. Two machines, the Hollander and Fourdrinier were the key to large-scale paper production and together had the greatest impact world-wide on the development of the paper industry.

The Hollander

The Hollander or beating engine was invented in Holland about 1650, brought to Britain about one hundred years later and appeared in Scotland early in the nineteenth century. It consisted of a drum fitted with bars of steel in which the rags, previously digested in boiling water, were lowered onto a bedplate fitted with blades that sheered and bruised the fibres to the required consistency. It replaced the preparation of pulp by hammering or stamping rags in a mortar

Hollander beaters in Carrongrove mill in 1915.

The Hollander, which beat raw material to a pulp, was in use from early in the nineteenth century

and could reduce twelve hundredweights of rags to pulp in the time it took 40 stampers or mortars to reduce one.

A Hollander machine was in use at Herbertshire Mill during the early nineteenth century as described by the Rev John Dempster in his account of the parish of Denny in the *New Statistical Account of Scotland*, published in 1845.

> *. . . As soon as the rags are cut by women across a scythe blade fixed into a table covered with wire-cloth, for the purpose of getting rid of the dust and sand, they are passed into the boiling house, where they are boiled for twelve hours; afterwards, they are washed, and broke into a pulp by an iron cistern, called a paper-engine, capable of holding one hundred weight of rags, which are beat by a roller with thirty-six steel bars, which turn on a plate in the bottom of the cistern. Five of these engines, of twenty steel bars, are kept constantly going night and day, requiring upwards of forty horse power to drive them and the other requisite machinery.*
>
> [*The New Statistical Account of Scotland*, vol 8, (Edinburgh, 1845) p128]

Making a pulp – these photos from the World War II period (note the women workers) show the same process of papermaking from the earliest days

(top) workers are loading newsprint into the pulping machine – straw was also used during the war years

pulp is delivered by chute into a revolving screen to remove contraries, the pulp passing through the perforations

the prepared pulp is collected prior to washing and further treatment

The Fourdrinier Machine

The Fourdrinier papermaking machine was invented in 1799 by Nicholas Louis Robert who worked at Francois Didot's mill at Essonnes in France. The machine, designed to produce a continuous web of finished paper, revolutionised the industry. It consisted of an endless loop of woven wire cloth onto which flowed a steady stream of pulp which was gently shaken to get rid of surplus water and then put through couch rolls and drying cylinders to finish as rolls of paper.

John Gamble, Didot's brother-in-law, took out an English patent on the machine in April 1801. It was first installed in England in 1804 at Frogmore Mill in Hertfordshire and in Scotland in July 1807 at Peterculter Mill in Aberdeen. Early versions of the machine were able to produce paper 24 inches wide and up to 15 yards long. The machine in Aberdeen was fully operational by 1811 and on 26 August 1812, the *Aberdeen Journal* was first published on machine-made paper.

Gamble, working with machine engineers, John Hall and his apprentice Bryan Donkin from their works at Dartford, went on to make further improvements to the machine increasing its speed from 20 to 34 feet per minute. By 1813, the Fourdrinier machine was commercially available to paper manufacturers throughout the country.

Herbertshire Mill followed suit installing a paper machine to compliment its beating engine thereby enhancing production levels and profit. It too is described in the Rev Dempster's account of Denny alongside another Fourdrinier invention for cutting paper.

> *. . . After the rags are broke in and bleached for twenty-four hours, they are beat into pulp or stuff ready for passing on to the paper-machine, perhaps one of the most complete pieces of machinery ever invented in this country; as, in one room of 60 feet in length, by 24 feet wide, one may see the stuff much resembling churned milk,*

'The Fourdrinier machine consisted of an endless loop of woven wire cloth onto which flowed a steady stream of pulp . . . then put through rollers and drying cylinders to finish as rolls of paper'

passing by means of a fine web of wire-cloth fifteen feet long into a series of rolls used in pressing out the water, and forming the paper into a firm body. It then passes into a set of cylinders heated by steam, from which it is reeled into rolls in a perfectly finished state, quite dry and pressed, ready for use. Six of the rolls are then put on to the cutting-machine,which cuts them into the sizes required. The cutting-machine is the invention of Messrs Fourdrinier of Hanley, Staffordshire, and patent. It is capable of cutting 144 sheets per minute of post or writing-paper. On an average, 26 cwt. of rags are cut per day in the rag-house, and 21cwt. of them beat into stuff, yielding an average of from 1600 to 1700 lbs. per day of twenty-four hours, as all the machinery is kept going night and day.

[The *New Statistical Account of Scotland* vol 8 (Edinburgh, 1845) p128]

Throughout the nineteenth century, paper machines were improved upon in terms of size, width, speed, efficiency and levels of production. As one machine was increased in size so each supporting one had to be adapted to keep pace with new developments.

Early Hollanders had vertical sides and flat bottoms which left pockets of pulp unbeaten so the shape of beater troughs were altered to curved sides. The width of Fourdrinier machines increased from 48 inches, common in the 1820s and 1830s, to 80 inches, as displayed by Bertrams at the 1862 London International Exhibition. Machines capable of reaching speeds of 100 feet per minute were also on display. By the 1900s, 150 and 160 inch wide machines were common, operating at speeds of 420-480 feet per minute.

Steam power was not essential to the installation of paper-making machines but steam itself had long been used to heat vats and warm drying lofts. A Watt steam engine was installed in the Devanha paper mill of Brown, Chalmers and Company at Craigbeg, Ferryhill, Aberdeen in 1803. The steam engine certainly contributed to the increased power needed to cope with expansion but it was the paper-making machine, mainly water-driven, that

Papermaking machine No 1 photographed in the 1890s at Carrongrove

Falkirk Museums P15759

No 2 papermaking machine photographed in 1930

changed the industry from unit sheet production to continuous production.

The Development of Engineering Companies

The mechanisation of the paper industry gave rise to a new industry in its own right, that of the manufacture of papermaking machinery. The local wright and joiner could no longer cope with the increasing complexity of machinery required. As wooden machines and parts were gradually being replaced by metal, specialist engineering firms such as John Hall from Dartford and Bertrams of Edinburgh emerged. From 1794-1807, John Hall supplied equipment to Scottish mills and trained engineers, including the Bertram brothers, who went on to construct and develop their own ideas.

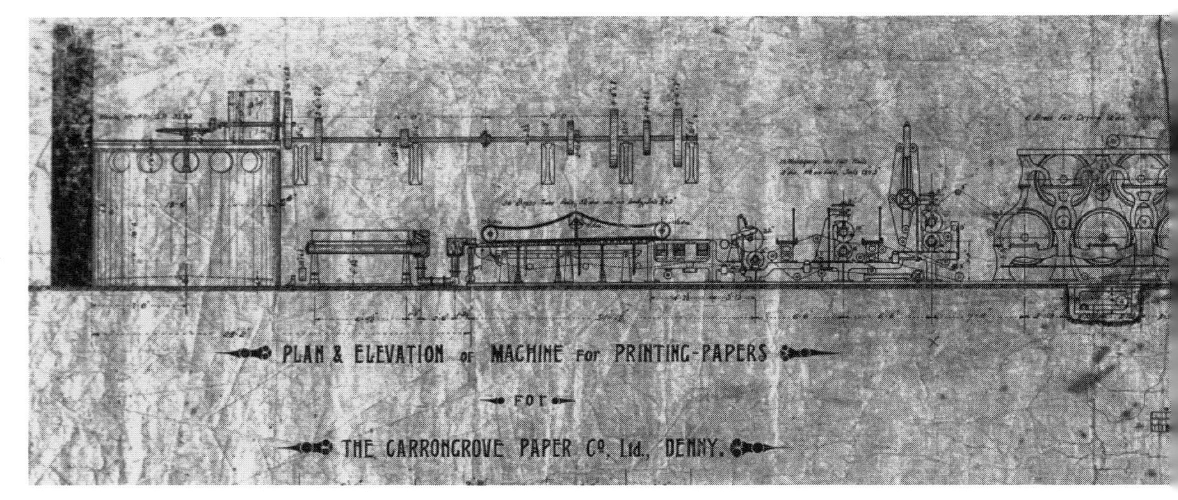

Courtesy of the National Archives of Scotland GD1/779/10

Bertrams Limited

George and William Bertram worked with their father at Springfield Mill, Polton, were apprenticed to John Hall at Dartford and returned to Edinburgh in 1821. William started an engineering works while George became a paper-maker at one of the mills on the Esk. He then joined his brother in manufacturing paper-making machines and developed St Catherine's Works in Sciennes, Edinburgh. Their younger brother, James started his own paper engineering firm in Edinburgh in 1845.Together, their two works, St Catherines and Leith Walk, earned a world-wide reputation for quality machinery.

The United Wire Works

Another key business in the development of the industry was the production of continuous woven wire to support the Fourdrinier machines which was supplied by the United Wire Works established in Granton in 1837 by William McMurray.

William started up as a wire worker in Glasgow in 1825. He moved to Edinburgh and set up himself and his brother, James as wire-cloth workers on Leith Walk. Thriving well into the twentieth century, they supplied wire cloth capable of withstanding the increased speeds and demands of modern machinery to Bertrams and most other major paper-machine manufacturers.

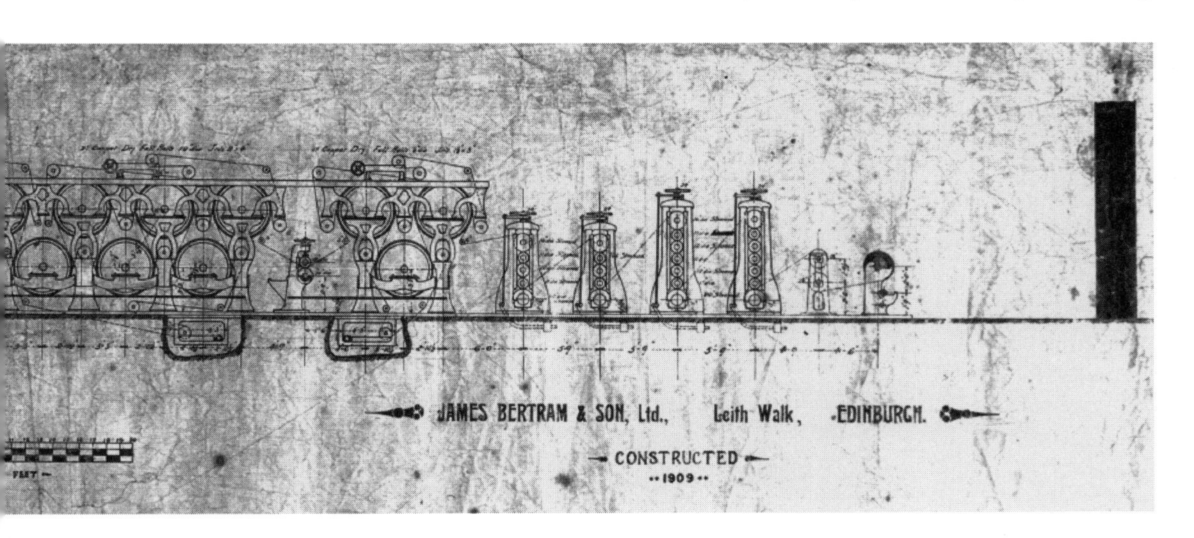

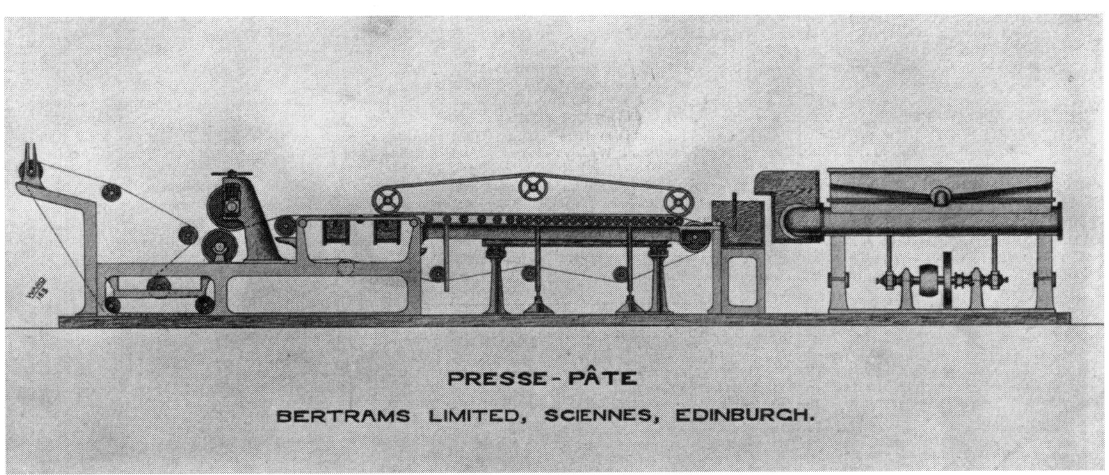

Bertrams of Sciennes, Edinburgh became one of the industry leaders in manufacturing papermaking machinery and supplied Carrongrove mill

Courtesy of the National Archives of Scotland RHP 50085

The Esparto Grass store shed at Carrongrove in 1920

Courtesy: Tom Stein

Esparto Grass and Wood Pulp

From the late nineteenth century, the search for new raw materials to supplement or replace the shortage of rags turned to the use of esparto grass and wood pulp.

Esparto grass grew in Spain and Africa. It was wiry and tough and an economical source of fibre for papermaking. It was first imported into England in bulk in 1851. Key patents were taken out by Thomas Routledge in 1856 and 1860 who was first to produce paper from esparto pulp at his Eynsham Mill in Oxfordshire. He received a medal for his esparto paper at the International Exhibition of 1862 which caused other papermakers to latch on to its potential.

This new fibre required different handling from rags.The grass, when delivered to the mills in huge bales, was broken open and sorted by women, put through a machine called a 'willow' to beat out the dust and dirt and discharged onto a conveyer belt where any roots and weeds were removed. The material was then fed into digesters where a mixture of lime and soda or potash was added and boiled for four to five hours. The residue caustic liquor, which then turned black, was drained into storage tanks for later treatment to remove as many of the chemicals as possible. The esparto was then taken from the digester to be washed and beaten in a type of Hollander called a 'potcher' to remove the soda and remaining organic matter. The pulp was then rinsed in a solution of carbonate or bicarbonate of soda or bleached to prepare it for final cleaning, drying and moulding into paper.

In 1861, nine mills in the Esk area made use of 148 tons of esparto in conjunction with 6,565 tons of rags. From then on the use of esparto spread quickly throughout Scotland especially where mills had easy access to harbours. By 1914, imports of esparto grass

Courtesy: Tom Stein

had risen to 183,114 tons. Scottish firms soon became specialists in this field producing esparto paper with a close texture and smooth finish that found a ready market in the printing industry.

The consumption of esparto was affected by the two World Wars when mills were forced to turn to straw as an alternative source of fibre. Faced with the growing challenge of wood pulp, used mainly for newsprint and less expensive gradings of wrappings and printings, uncompetitive prices and the economic pressure of costly effluent treatments to counter river pollution, plants closed down.

Carrongrove in the 1800s

THE importance of plentiful supplies of good quality water to papermaking at Carron Grove was evident in the early years of the nineteenth century. Records show Robert Weir and Gavin Glennie were known to one another in respect of a lengthy court case which Robert Weir pursued through the Sheriff Court and Court of Session from 1830-1832, over maintenance of the mill lade on the Carron and the corresponding supply of water to his and neighbouring mills.

Robert Weir owned two mills, one on each side of Gavin Glennie's, a corn mill upstream and his paper mill downstream. All three were served by water passing through the same mill lade. At first, Weir's complaint against Glennie was that water passing from the latter's mill was so foul and slow moving as to prevent the proper function of his paper mill. As the case dragged on, Glennie and other mill owners were accused by Weir of trespassing on his private road running along the north side of the canal at Tamaree and interfering with the dam sluice causing the canal to overflow and damage sections of the bank on his property.

In his defence, Glennie's lawyers placed the blame on Weir himself for failing to maintain the canal banks. Their version of events on the day in question was that on inspecting the sluice following a break in the usual supply of water, they found

> *"an old dike or rickle" on Robert Weir's land had given way and about the equivalent of 3 cart loads of stones had fallen into the canal to choke it, impede the water which then overflowed its banks.*
> **[NAS: Court of Session papers: CS230/W/8/29]**

The case eventually came to a close on 16 February 1832 with the Court failing to find in either favour or awarding expenses to either side. Both parties were granted right of access along the canal bank and were to agree to keep it in good repair.

Access to a regular supply of Carron water remained a problem for Robert Weir and others for years to come. On 30 May 1836, he was one of several signatories to a letter representing the *Committee of Proprietors for the Better Supply of Water to the Mills on the River Carron* addressed to William Forbes of Callendar, their local MP, requesting a financial contribution towards the cost of securing such a supply. [7] Forbes himself owned three mills on the river so we can only assume the signatories were hopeful of a favourable result.

Carron Grove Mill was put up for sale in 1833 and bought by Robert Lusk.

In 1834, Robert Weir leased Herbertshire Mill to a new tenant, Andrew Duncan, whose intention to install new machinery at a cost of £2000 to £3000 and add new houses to the premises is recorded in a rental and valuation for Herbertshire Estate, dated 18 June 1835. [8]

ON THURSDAY night last, a young man by the name of More had a serious accident at Mr Lusk's Paper Mill at Carrongrove. The unfortunate man had his arm caught in a calendering machine and was so badly shattered that it had to be amputated.

Report from the *Stirling Journal* of 17 February 1837

Again, drawing on the Rev John Dempster's account of the parish of Denny in the *New Statistical Account of Scotland*, 1845, he includes a detailed description of business at this time in both Herbertshire and Carron Grove Mills. Whereas Herbertshire, the larger mill, concentrated on the manufacture of paper from rags, Carron Grove used rope for the production of millboards and some grades of coarse papers. Andrew Duncan employed a total of seventy people, twenty men and fifty women with the help of paper-making machinery.

> *The wages are paid monthly; on an average, 15s. per week for the men, and 5s. for the women: besides these, 2 men and 4 horses are constantly employed carting rags and coals and carting the paper to Grangemouth for shipment to London. . . The duty paid every six weeks averages L.320; the wages every month, L.100; carting, and other carriages, L.40. The water-wheel for driving the paper engines is 24 feet diameter, and fully 12 feet wide, all*

> *iron, and weighs 33 tons. Another small wheel is used for driving the paper-machine, 22 feet diameter, and 18 inches wide. The works are lighted with gas, and four tons of coals are used daily.*
> **[*The New Statistical Account of Scotland* vol 8 (Edinburgh, 1845) p128-129]**

At Carron Grove Mill, Robert Lusk employed twenty-five workers,

> *. . . 15 men, 2 women, 2 lads, and 6 boys; wages are paid every fortnight, and average about L.27 every fortnight. The materials used are almost exclusively old tarred ropes, of which fully one ton, on an average, is used daily. No rags are made use of in this manufactory. The goods manufactured are almost exclusively millboards, which are used for the boards of books, of which from four to five tons per week are manufactured. Sometimes a little coarse paper is made, used for sheathing ships and other purposes, as also some large coarse millboards, used by engineers, for making steam-joints tight. The mill is lighted by gas, and the manufactured goods dried by steam and heated air. The excise duty paid is from L.300 to L.400 every six weeks.*
> **[*The New Statistical Account of Scotland* vol 8 (Edinburgh, 1845) p127]**

'a common concern among many nineteenth century mill operators was the burden of excise duty on paper and board'

Robert Lusk certainly faced his share of difficulties in running the mill involving both tax and fire. A common concern among many nineteenth century mill operators was the burden of excise duty on paper and board. On 26 July 1836, Robert Lusk also wrote to his local MP, William Forbes of Callendar, asking for his support in his claim against the Chancellor of the Exchequer and the government's proposed fifty per cent reduction in paper duty that year. Lusk wanted existing laws to stand which guaranteed a discriminating duty in favour of pasteboards and millboards which, in his opinion, were essential to the very existence of his trade. He wanted compensation for losses incurred as a result of this new legislation. He wrote in his letter:

First, I desire to obtain an audience of the Chancellor of the Exchequer in order to lay before him the true nature of my case which he has either misunderstood or evaded in all his replies on the subject, and to meet the objections he makes against giving any compensation. . .

2. . . I wish to be able to bring the matter under the notice of both Houses of Parliament, especially before the House of Lords, where there is more apparent respect for the legal rights of the subject, and I wish you to . . . direct me how I ought to proceed, or to introduce me to those likely to take the matter in hand.

[Falkirk Archives: Forbes of Callendar papers: GD171/1239/26]

Sadly, we have no record of the outcome of his case though years later, in June 1853, Andrew Duncan again sought the help of William Forbes to lobby parliament against a proposed five per cent increase in paper duties. In his letter, he wrote:

. . . I hope you will be able to get this redressed. . . I can assure you that you will do the trade a very great good.

[Falkirk Archives: Forbes of Callendar papers: GD171/1359/41]

'. . . I hope you will be able to get this redressed. . . I can assure you that you will do the trade a very great good'

On 8 January 1841, according to a report in the *Stirling Journal*, Carron Grove Mill was damaged by fire with the loss of £2000 worth of paper, machinery and twenty jobs. [9] Come 1847, the mill changed hands to Robert McRobie who continued manufacturing pasteboard there until his death in 1850 when his brother John took over. Five years later, John McRobie sold the mill to Robert Weir giving Weir ownership of both Herbertshire and Carron Grove Mills.

In 1858, Weir brought in John Luke as manager of Carron Grove. Luke was an experienced and accomplished paper maker who had learnt his trade at Airthrey Paper Mill in Bridge of Allan and whose sons, in later years, went on to found Anchor Paper Works and the Vale Paper Company in Denny. [10]

Weir sold Herbertshire Mill to Andrew Duncan in 1860 after Duncan's many years successfully running the mill as his tenant. The following year, Duncan started to build Glencarron House the property housing the offices of Carrongrove Board Mill today. On 10 October 1862, the *Stirling Journal* reported the completion of his house.

> *A considerable number of fine buildings have been erected in this town and neighbourhood during the past few months. The splendid new mansion house of Andrew Duncan, Esq., paper manufacturer is now ready for entry. The architectural appearance, general outline and dimensions of this building will, we are convinced, stand a favourable comparison with many of the finest in Stirlingshire. Its situation is excellent, being seated on the banks of the Carron in close conjunction with the magnificent Carron Glen.*

Sadly, Andrew Duncan did not live long to enjoy his new home dying less than a year after its completion on 27 August 1863, aged only 53. His obituary in the *Stirling Journal* of 5 September 1863 acknowledged his success as a paper manufacturer.

> *One of the largest funerals ever to have taken place in the neighbourhood took place this week, when the remains of Mr Andrew Duncan were conveyed from his residence, Glencarron, and interred in the parish Churchyard. In 1834, Mr Duncan became one of the lessees of the Herbertshire Paper Works, then in a most ruinous condition. He devoted all his energies to its development and with such success that in short time the superiority of his manufacture of paper was recognised in England as well as the Scottish markets. He finally became proprietor of Herbertshire Paper Works, purchased the grounds*

and erected one of the most beautiful, if not extensive, mansions in the county.

[T Clapperton ***History of Carrongrove Paper Mill*** **Appendix A]**

Andrew died a wealthy businessman leaving his wife, Janet Douglas and their five children, Alexander, Walter, James, Elizabeth and Janet, and an estate valued at £5,653 19s 2d. His will lists his assets in great detail including cash, household furniture, silver plate, china, books, pictures, wines, a gold watch and jewellery, bed and table linen, clothes, horses and carriages, garden tools and implements and paper mill utensils. It also confirms his intention that his sons, Alexander and Walter should follow him into the paper making industry. [11]

Alexander took over the running of Herbertshire Mill and set about modernising and developing the firm. In 1869, the company installed a Bertrams No 50 84 inch paper making machine and by 1872, employing 200 people, was capable of producing 16 tons of paper per week and 832 tons of rag and esparto printing papers annually. [12] In 1879, the company took over the neighbouring Stoneywood woollen and dyewood mill, converted it into a second papermill and installed an 84 inch Bertrams Machine the following year. Alexander Duncan and Sons operated Stoneywood Mill until 1886 when it passed into the hands of John Collins, a paper manufacturer from Glasgow.

In 1893, John's father, Sir William Collins, bought Herbertshire Mill from Alexander Duncan and Sons and placed John in charge giving the latter responsibility for both Herbertshire and Stoneywood Mills. John died in 1895 but the company John Collins Ltd continued, providing work for 250 employees producing laminated papers and envelopes with an output of 80 tonnes of paper per week.

Robert Weir himself died in 1866 at the age of 83. His life-long achievements were duly noted in his obituary in the *Stirling Journal* of 30 November.

> *For 60 years the name of Robert Weir has been eminent among the stationers of Glasgow. A native of Edinburgh, he came early to Glasgow to follow his calling and began life as a stationer at the age of 21. At the end of 20 years of successful industry, he added to his Glasgow business, Herbertshire Paper Mills, and shortly afterwards, he established a stationery warehouse in Montreal. Later on in life, he sub-let Herbertshire Mill to its present proprietors and bought the adjoining estate of Randolph Hill. . . Carrongrove Paper Works and other property. A year or two ago, he ceased to be a partner in his Montreal house, and as late as three months ago, let Carrongrove Paper Works.*

'Plant and Machinery therein. . . Water Wheel and Waterfall, Beating Engines. . .'

At the time of Weir's death, Carron Grove Mill was sold as a going concern. Press notices in the *Stirling Journal* for 9 August 1867, 13 March and 10 April 1868 record its sale among his other assets including a woollen and meal mill, Tamaree Cottage, Randolphill Mansion House and Fankerton Cottages. The sale notice for Carron Grove Mill itself, 9 August 1867, included

> *Plant and Machinery therein. . .Water Wheel and Waterfall, Beating Engines, Rope Cutter, Steam Boiler, Steam Drying Pipes, Calendars, Dry and Wet Presses, Cutting Machines, Yankee Machine etc.*

Carron Grove Paper Mill and Randolphill House passed into the hands of John Miller, founder of John Miller Ltd., Printers, St George's Road, Glasgow, who set about improving and modernising the mill. He bought the neighbouring Baty's Wool Mill and rearranged the entire Carrongrove layout replacing the old iron water wheel (24 feet in diameter, 12 feet wide and weighing 23 tons) with water driven turbines and installed a Bertrams paper machine with an 81inch wire and eleven 4 feet drying cylinders. [13]

In 1875 John Miller returned to the stationery business and sold Carron Grove Mill to Messrs. Plummer and Henderson from

Edinburgh. Under their management, the mill went on to produce 18 tons of newsprint and low grade printings per week paying weekly wages of 25 shillings for beatermen and 27s 6d for machinemen. [14] Within two years, they gave up the mill and returned to Edinburgh, Plummer back to a paper agency and Henderson to start Bonnington Paper Mill in Leith which he owned until his death.

A new company, Carrongrove Paper Company Ltd was formed in 1877 and James Johnston, who had been manager of James Brown's Esk Mills in Penicuik, was appointed manager. He ran the company for eight years but his success was cut short by his untimely death from drowning reported in the *Stirling Journal* on 22 June 1885.

> *Early on Friday morning, James Johnston, partner in the firm of Carrongrove Paper Co., Denny and manager of the mill there, was found drowned in the lade that supplies the works and is fed by the River Carron. Mr Johnston, as he frequently did, paid a visit to the mill towards midnight on Thursday and on making his way to his residence at Randolphill, it is supposed that he stumbled into the lade, which it is necessary to cross.*

'A new company, Carrongrove Paper Company Ltd was formed in 1877'

Johnston was succeeded by William Walker, who came from Kevock Mill in Lasswade, and stayed for two years. During his time in office, he supervised the installation of a steam boiler plant, a main drive engine, plate glazing calenders and a second paper machine. [15]

Some years later, around 1895, an article published on the development of the paper industry in Denny described its success, in particular the achievements of Carrongrove and its then manager, George Johnston from Culter, nephew of James Johnston and successor to William Walker.

> *The valley of the River Carron has now become a very important centre of the paper manufacturing industry, which is fast*

Falkirk Museums P13115

Carrongrove at the end of the nineteenth century – note the edifice of Carrongrove House built by former tenant and then mill owner, Andrew Duncan in 1862

displacing the woollen factories that formerly congregated hereabouts. . . The two industries carried on simultaneously until 1868, when it was decided to drop the woollen trade and take up the manufacture of white paper. The mills were then thoroughly overhauled and equipped with the latest and most improved machinery adapted to the paper industry. In 1877. . . the Carrongrove Paper Company acquired the business and under their auspices, it has been developed with remarkable rapidity. In fact, during the last sixteen years, they have doubled the productive capacity of the mills and the weekly output is now averaging 80-tons.

During the past few years, the company have devoted their attention to several specialities in the way of papers now so popular, for which they are noted throughout the trade. Among these may be mentioned their well-known tinted writing papers. . . cartridge and drawing papers, papers for enamelling and cream laid papers for account books. They also produce printing papers and glazed envelope paper, as well as other varieties. The works cover an area of 27 acres and are admirably arranged and organised throughout. Upwards of 250 hands are employed. A siding from the Caledonian Railway runs right into the works, thus affording excellent facilities for transport purposes. . . water

A team of labourers with horse and cart extend the premises to house new papermaking machinery around 1910

power is largely utilised in driving the machinery in the mills. . . They have. . . built up an extensive and influential connection not only at home but in the Colonies and America and upon the Continent. The manager of the Company is Mr George Johnston, a gentleman of long experience in the trade, to whose energetic supervision, the success of the enterprise is in no small measure due.

[T Clapperton *History of Carrongrove Paper Mill* p11-12]

George Johnston managed Carrongrove Paper Company from 1887 to 1898 during which time the mill became a major contributor to the papermaking industry in Scotland. As part of his programme of improvements, a new evaporator was installed to concentrate the spent liquor which was discharged from the esparto grass boiling process. Previously allowed to flow into the River Carron, after leaving the evaporator, this now concentrated liquor had a tar-like appearance and was fed into the back of a rotary furnace. The resulting grey ash was a crude form of sodium carbonate which was reused in the manufacture of caustic soda. According to the *Stirling Journal*, October 1894, the installation work was completed under Thomas Marshall, foreman engineer, at a cost of £2,000 following recommendations from the Eastern District Commissioners of the County Council to Carrongrove to reduce levels of river pollution.[16]

George Johnston also installed new paper making machines supplied by Bertrams of Edinburgh. Entries in Bertram's illustrated catalogues for their St Katherine's Works in Sciennes contain testimonials from Carrongrove praising the quality of their machines and standard of service.

> *Having long experience of nearly all kinds of machinery by you for use in the manufacture of fine papers, we have great satisfaction in testifying to the substantial, accurate, and well-finished machines you turn out. We would more especially refer to your make of Paper-Making Machines with their necessary adjunct of Strainers (either in the form of 'Revolvers' or your 'Patent Flats'), which in our experience are unequalled by any other maker.*
>
> **[NAS: Bertrams Ltd Sciennes: GD1/779/10, p46-47]**

> *We have much pleasure in stating that the alterations of our breaking and beating plant with new shafting and drives that you carried out for us in July last* (1897) *have proved very satisfactory.*
>
> **[NAS: Bertrams Ltd Sciennes: GD1/779/10, p32-33]**

'according to Bertrams, the key to the successful glazing of paper lay in efficient damping'

According to Bertrams, the key to the successful glazing of paper lay in efficient damping, using either rolling or friction calenders. Carrongrove thus made use of Bertrams Milne's Patent 'Victory' Damping Machine, most efficient *'for imparting any amount of moisture necessary to paper or cloth'.* **[NAS: Bertrams Ltd Sciennes: GD1/779/10, p84-85]**

Bertrams benefitted not only from Carrongrove's custom but also from the development of a purification stockboard invented by James Blaine, a foreman papermaker at Carrongrove. The patent for this invention passed into Bertrams' hands who then introduced it into leading paper mills throughout the country.

Twentieth Century Developments

THE paper industry remained generally prosperous up to the First World War with Britain being the third most important paper producer in the world after America and Germany. British mills came under government control during both World Wars and up to 1956 when the last import controls on raw materials and paper were removed.

Rags virtually disappeared from papermaking after the Second World War as did the use of Hollanders and other beating machines. The twin wire or double machine was invented in 1923, alongside the development of super calenders and new machines incorporating plastics and other new components. A faster more adaptable board machine, the Inverform, was built in the 1950s by R J Thomas at St Anne's Board Mill in Bristol and developed further by Thames Board Mills in 1963. Fourdrinier machines were replaced with others capable of speeds of 960 feet per minute for paper production and 4,800 feet per minute for newsprint production.

'rags virtually disappeared from papermaking after the Second World War'

The first computer-controlled paper machines were installed in Wolvercote Mill and later Grove Mill in Disley in 1966. These could monitor the amount and quality of paper made and record machine speeds and performance. Today, the influence of the computer has been extended to the entire mill operation from ordering stocks of raw materials and chemicals, to resetting machines to run different grades of paper on demand and despatching paper orders to all corners of the world.

Despite a decrease in the number of mills and employees and little opportunity to modernise machinery during the immediate post-war years, output continued to rise. Demand rose from 30lbs of paper per head in 1907 to 328lbs by 1986.

Great advances were made in recycling waste paper to make cardboard, packaging and board and in the production of light weight coated paper for stationery, cards, colour magazines and mail order catalogues.

The integrated mill was introduced in the 1960s taking in the entire process from the delivery of the tree at one end to stockpiles of paper at the other. Wiggins Teape established a chemical pulp and paper mill at Corpach near Fort William in 1963 to use local supplies of wood for the production of high quality paper and card. Both pulp and paper mills were on stream by 1966 but due to the high cost and irregular supply of the timber, the pulp mill was forced to close. A second integrated mill, The Caledonian, south of Glasgow, was opened in the late 1980s to convert Scottish spruce into light-weight coated paper suitable for glossy magazines.

Serious difficulties were experienced during the 1970s when imports outstripped domestic production and demand for paper dropped. A combination of entry into the Common Market in 1973, which opened up wider competition for the industry, huge rises in world oil prices, inflation, a drop in consumer confidence and slow economic growth added to this slump in trade.

The greater availability of energy sources at competitive prices, the development of computers, new technology and recycling plants to treat and dispose of effluents have helped reverse this trend. Confidence returned in the 1990s and promoted the commissioning of new machines.

Courtesy: Tom Stein

A close-up of Carrongrove House. The Model T is thought to have been owned by mill manager William Wallace. From 1898 when William Wallace took charge, the mill was continuously managed by father and then son, Morgan Wallace until 1967 followed by his younger brother Alen

Carrongrove in the Twentieth Century

WHEN General Manager George Johnstone was forced to retire from ill-health in 1898 and died shortly afterwards at the early age of 41, he was replaced at Carrongrove by William M Wallace. For most of the next half century William Wallace took responsibility for further major developments within the company consolidating the production of top quality paper and overseeing the purchase of Herbertshire and Stoneywood Mills from John Collins Ltd.

Wallace was appointed to the Board of Directors in 1903 and then became Managing Director of Carrongrove Paper Company in 1909. His plans for expansion received a set back when fire broke out at the mill in 1905 destroying the engine-house. The *Falkirk Mail* of 9 September reported the incident as follows:

A disastrous fire occurred in the Carrongrove Paper Works, Denny on Thursday night. The outbreak was discovered in the floor of the engine-house by the engineman between 8 and 9 o'clock. The mill fire brigade soon got to work, but not withstanding their efforts with a plentiful supply of water, the engine house was quickly destroyed and a huge and expensive 600 H.P. steam engine was much damaged. At present the amount of damage cannot be accurately stated, but it must amount to over £2000. The mill, which employs about 200 hands was thrown idle yesterday and today but work is expected to resume on Monday. The cause of the fire is unknown.

'of considerable moment to the commercial life of the district'

A full recovery was made and the following year Carrongrove Paper Company put in a bid to buy the neighbouring Herbertshire Mill. The *Falkirk Mail*, 10 November 1906, reported the meeting of the shareholders of Stoneywood and Herbertshire Paper Company as being

. . . of considerable moment to the commercial life of the district. An offer by the Carrongrove Paper Co. for the purchase of Herbertshire Mill and grounds and the right of the water supply has been placed on the agenda and the managing directors recommend the acceptance of the offer. . . Indeed, the sale of it is said to be practically completed. The output of paper, which is of duplex type at Herbertshire reaches something like 35 tons a week and is an excellent paying concern. In the hands of such an enterprising Company such as Carrongrove, the mill would undergo much improvement and the Company would have greater scope for extensions than they can at present readily command. The offer made by the Carrongrove Paper Co. is stated at £14,420.

[T Clapperton *History of Carrongrove Paper Mill* Appendix B]

Herbertshire Mill was bought by Carrongrove. On 17 November 1906, the result of the shareholders meeting in favour of the sale was printed in the local newspaper. The editorial concluded. . . *'the employees in these parts are now on the tip-toe of anxiety as to the fate of the old mill'.*

Herbertshire Mill closed on 12 January 1908 with the loss of eighty to ninety jobs. William Wallace's plans for Herbertshire involved dismantling the mill and clearing the site to make way for extensions to Carrongrove, the installation of a 112' paper machine, one of the largest in the country, and the conversion of Glencarron House into offices for company use.

Demolition began on 1 February with Herbertshire's 168 feet high chimney as reported by the *Falkirk Mail*:

> *. . . notwithstanding the rough and inclement weather, a large crowd assembled at Herbertshire Paper Mill to watch the demolition of the tall chimney stalk. Men had been engaged for some days loosening the foundation and so directing its fall to a given spot. A charge of dynamite was set to cause its upheaval and a current of electricity did the rest. The electric button was, at a given signal, touched by Mrs W M Wallace of Randolphill, and the stalk quickly fell in a heap where it was intended.*

'the employees in these parts are now on the tip-toe of anxiety as to the fate of the old mill'

By 1909, trees had been cleared from Glencarron grounds to make way for electric cables to power Carrongrove Works, the No 2 paper machine was remodelled, and a third, the 112' sheet paper machine, was installed. A fourth paper machine and more auxiliary plant were installed in 1912 and Wallace took up residence in the new Randolphill House, the original having been demolished and completely rebuilt.

In 1906 Carrongrove bought over Herbertshire Mill which had been operating two machines, a 76 inch and a 57 inch, in the production of machine-glazed papers. Stoneywood Mill remained in production until its closure in 1984.

The First World War held up further development at

Carrongrove but upgrading and expansion continued throughout the 1920s despite two changes of ownership. The company was renamed Esparto Paper Mills Ltd in 1922 and was then taken over by the Inveresk Group in 1924 who restored the original name of Carrongrove Paper Mill.

Business continued with Bertrams who regularly supplied and serviced Carrongrove machinery. Bertrams' Immediate Order Books are full of Carrongrove job entries, for example:

'upgrading and expansion continued throughout the 1920s despite two changes of ownership'

3 August 1921
. . . one new driving shaft for 6 feet Simplex Strainer on No.1 Machine.

[NAS:Immediate Order Book, 1920-24, Bertrams Sciennes Ltd:GD419/14/1 p105]

4 April 1922
. . . one set of 5 Chiller Calender Rolls to come here to be buffed to our best standard to a dead joint with a pressure of 20 cwts on each end; ex No.3 Machine.

[NAS:Immediate Order Book, 1920-24, Bertrams Sciennes Ltd:GD419/14/1 p218]

In 1923, transport facilities were brought up to date with the laying down of railway connections to the various mill departments. Work started on building a coating factory for the production of high class coated papers made from best Esparto quality paper made on the premises. Orders went out to Bertram Sciennes for Super Calendars Nos.980/981. By March 1925, four coating machines were in operation. Weekly output grew to 200 tons of plain paper and 230 tons of coated, which were both in steady demand at home and abroad.

William Wallace died in 1946, a well-respected company man and member of the community. On the occasion of his marriage in 1901, he had received a black marble timepiece and side ornaments fitted with electric lights from the mill employees. [17]

William's son, Morgan, took control of the mill until his retiral in 1967, to be succeeded by his younger brother, Alen who renewed the steam and boiler plant, added a new engineering shop and installated another new paper machine.

According to the *Burgh of Denny and Dunipace Official Guide, 1965*, the major employers in the town at this time were the paper mills, board mills and iron foundries. [18] Under the influence of the Wallace family, Carrongrove enjoyed seventy years of steady growth. Production levels rose from 80 tons of paper per week to a total of 370 tons culminating in a thirty acre site employing over 600 workers at Carrongrove, the Coating Plant and Stoneywood Mill.

'Carrongrove enjoyed seventy years of steady growth with production levels rising from 80 tons of paper per week to a combined total of 370 tons'

The mill started to lose money during the 1970s as a result of the world-wide recession. A succession of managers, J B Henderson, Jim Donald and Gordon Hall were forced to rationalise and modernise to survive growing competition from abroad. Paper Machines Nos 1 and 2 were closed down after a hundred years service. Between 1971-1977, with the emphasis on automation, instrument control and recording in all departments, staff numbers dropped to 377 and Machine Nos 3 and 4 were rebuilt and extended to increase production by one hundred per cent.

The Inveresk Group relinquished control of the company in 1981 selling Carrongrove to the Georgia Pacific Corporation. The new owners set about a £3.5 million capital expenditure programme to consolidate and revitalise the business. Stoneywood Mill was closed in 1984. New machinery and technology was introduced at Carrongrove to produce quality board for game cards, greeting cards, cosmetic packaging, record, video and CD sleeves, all items for a changing market. By 1989, staff numbers had fallen to 230 while mill capacity rose from 23,000 to 35,000 tonnes per year.

(left) The steam boiler plant installed in 1960 designed to produce 70,000lb of steam pressure /hour – this produced a power supply to the papermaking plant and to workers' houses and was used in papermaking

(top) Water turbine used as a back-up power supply

(middle) An on-site 5000kw turbo-alternator for back-up power

(bottom) General Manager William Wallace inspecting foundations for a warehouse extension in 1937

Hollander beaters making pulp for Nos 3 and 4 machines around 1940 – four beaters were needed to feed each papermaking machine

Falkirk Museums P427

Papermaking machine No 4 installed in 1912

No 4 machine operated by Jimmy Scott in the late 1950s

George Scott fitting a dandy on No 4 machine, late 1960s.

The dandy laid a watermark on the paper

(below) No 4 machine working with the dandy in place in the late 1960s – on the left is Michael Keegan, now a production supervisor with some 35 years service in the mill

(left) In the coating factory where finishing coat is put on the paper. Adam Bradbury is operating the super calendar

Two views of the cutting house in 1926 (above) and in the 1940s (below). Note the number of women employed

(top left)
Counting and packing in the 1920s

(bottom left)
Overhauling, inspecting and counting in the 1930s

(top right)
Samples being taken from every order

(bottom right)
An overseas order packed and wrapped in hessian

TAMAREE
CARRONGROVE
PAPER Co. LTD.
DENNY.
MWG 229
The
Carrongrove
Paper Coy. Ltd.
Denny.
CARRONGROVE
ERF
TMS 864

Transport for the disposal of flue dust and ash generated by coal burning (top and middle) and for paper delivery (bottom)

(top right)
A direct railway line into the Carrongrove plant in 1923 brought coal in wagons with the Carrongrove livery. Wood pulp and esparto grass were also delivered by rail, hundreds of tons at a time

(bottom right)
transport costs had to be keen as the letter from Denny station to the mill manager shows

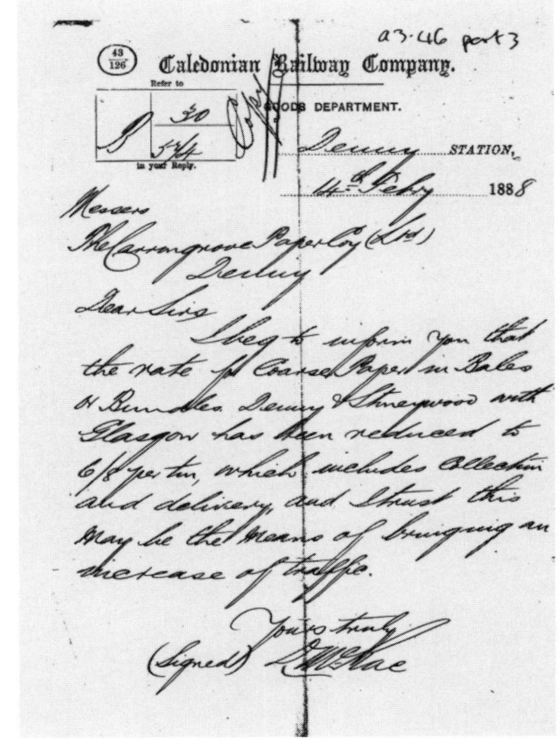

a3.46 part 3

43/126 Caledonian Railway Company.

Refer to 30/574 in your Reply.

GOODS DEPARTMENT.

Denny STATION,

14th Feby 1888

Messrs
The Carrongrove Paper Coy (Ltd)
Denny

Dear Sirs

I beg to inform you that the rate for Coarse Paper in Bales or Bundles Denny & Stoneywood with Glasgow has been reduced to 6/8 per ton, which includes Collection and delivery, and I trust this may be the means of bringing an increase of traffic.

Yours truly
(Signed) L McRae

List of papers made at Carrongrove up to the 1970s

Coated
Superfine Art, One & Two sided
Magazine Art, One & Two sided
Matt and Dull finish Art
Art Boards, One & Two sided
Chromos
Bright and Dull Enamels

produced on 4 coating machines – two 72" & two 78"

Plain Papers
TS & ES Cream Laid & Wove watermarked when required
Fine Printings
Featherweight
Super Finished Printings
Vellum Wove & Laid
Litho Printings
Antique Laid and Wove Printing
Tinted Writings
Azures
Offset Litho Cartridges
Deckle Edge Papers
Tinted Pulp Boards
Envelope Paper
Photogravure Printing
Enamellings
Duplex (Tint & White)
Duplicate Papers
Music & Plate Papers
Gummings
Etching Boards
Sensitized Cheque Paper
Drawing Cartridge
Pastings
White & Coloured Blottings
White Pulp Boards
Smooth Cartridge
Imitation Art
Parchment Wove

produced on 4 paper making machines – 75", 80", 90" & 92"

Before the concentration at Carrongrove on producing paperboard, many specialist papers were made, including numerous grades of papers, papers for the Crown Agents and for colonial markets.

This is a Shaw Ruling Machine in operation sometime in the 1950s.

Ruled paper was made at the mill for ledgers and other stationery products. The operator is JImmy Blaikie

McCutcheon Stirling Collection

Campaigning for the vote on the Denny – Stirling road between Quarter Entrance and Wellsfield Farm, 20th September 1884.

The banner on the front carriage reads 'Employees of John Collins, Papermaker, Denny'.

All the participants are wearing a white cap printed with the word, 'Franchise'

Community Life among the Mills

BY the end of the nineteenth century, Denny was a thriving industrial town, supported among other industries by three paper mills, Carrongrove, Herbertshire and Stoneywood. Mill workers put in long hours and worked hard but there was still time for a shared interest in political and social affairs.

Combinations or unions were not in great evidence among Scottish mill workers during the eighteenth century but certainly developed alongside mutual benefit and benevolent societies during the late nineteenth. In the lead up to the third Reform Act of 1884 which increased the number of Scottish constituencies from sixty to seventy-two and went a considerable way to giving all men the vote in Scotland, Denny paperworkers took part in their own franchise demonstration on 20 September. The photograph opposite shows mill workers from Carrongrove and Collins Mills on four horse-drawn carts decorated with paper crowns and banners on the Denny to Stirling Road between Quarter Entrance and Wellsfield Farm. The march was led by William Lithgow on horseback, dressed in fox skin and cocked hat, carrying a sword and shield. While members of the paper strainers supported a banner reading, *'We'll strain the House of Lords to the very dregs'*, supporters from Carrongrove rallied under the words:

We working men of Carrongrove
Our banners here display
To root out Tory Lords
We'll fight and win the day.

On 9 June 1894, a meeting was held in Denny Public Hall by local millworkers to discuss the amalgamation of their two existing branches of the National Union of Papermill Workers in the town. The motion to dissolve the two branches to form one called 'The Water of Carron Branch' was passed with agreement. [19]

GENERAL RULES

TO BE OBSERVED AT

STONEYWOOD PAPER MILL

1. The Working Hours for all, except Shiftmen, shall be from 6 a.m. to 6 p.m. Breakfast Hour, from 9 a.m. to 10 a.m.; and Dinner Hour, from 2 p.m. to 3 p.m. On Saturdays, Working Hours, from 6 a.m. to 2 p.m.; Breakfast Hour, from 9 a.m. to 10 a.m.

2. The Working Hours for Shiftmen shall be from 6 a.m. to 6 p.m.; commencing on Monday morning at 6 a.m., but terminating at 11 p.m. on Saturday.

3. All Workers are required to attend at the Mill at any time, when requested to do so.

4. No Shiftman is allowed to leave his work until relieved by the person whose duty it is to do so, or by some one appointed for the purpose.

5. All Workers, on entering the Mill, shall receive a Ticket at the Gatehouse, and hand it in on leaving. Each worker must lift and hand in his or her own Ticket.

6. Time-workers coming in from five to ten minutes late, shall have one Quarter-hour deducted from their time; and from ten to fifteen minutes late, one Half-hour deducted from their time. Workers more than fifteen minutes late may not enter the Mill until after the following interval.

7. A fortnight's warning must be given by workers before leaving, otherwise all wages due at the time shall be forfeited.

8. A fortnight's warning to workers may be dispensed with, and immediate dismissal, with forfeiture of all wages due at the time, may follow the undermentioned offences:—

(1.) Failing to attend at the Mill when requested to do so, except in cases of sickness.

(2.) Being under the influence of intoxicating liquor while on duty; failing to attend when required, owing to the influence of intoxicating liquors; or bringing intoxicating liquors into the Mill.

(3.) Damaging or wasting the Mill property, machinery, tools, paper, stores, &c. In addition to the above punishment, workers shall be liable for the amount of damage or waste committed.

9. Window glass broken through carelessness will be charged against the workers in department.

10. No Worker shall leave the department in which he or she is employed, unless in discharge of duty.

11. No Worker allowed to communicate with Strangers during working hours, or to bring Strangers into the Mill without first asking and obtaining permission from the Manager.

12. No smoking allowed within the Mill premises.

13. No carelessness or laziness at work.

14. No swearing or unruly conduct.

15. No Worker shall enter or leave the Mill premises except by the gate appointed for that purpose.

JOHN COLLINS.

NOTE.

Careless, lazy, or drunken workers are not wanted at Stoneywood Paper Mill. Good, careful, and honest workers may expect to be liberally dealt with.

August 12th, 1887.

John Mathie, Printer, Denny.

Courtesy: Tom Stein

On the platform at a company sports day in 1928

Back Row (left to right)
J Mackenzie, R Morland, A Moodie, A Brown, G Gauld, J Robertson, J Herd, R Sneddon

Front Row (left to right)
G Denholm, – Melvin or Melville (farmer), J Ferguson, Mrs W M Wallace, G Gauld, W M Wallace, W Wyness, – Cunningham

Kneeling
G Scott

Cooperation between mills, managers and workers extended into leisure time. Press reports in the *Stirling Journal* and *Falkirk Mail* give brief accounts of millworkers' sportsdays and annual summer excursions to a popular destinations such as Culross, Ayr, Rothesay and Portobello among others.

> *8 July 1893*
> *Last Friday the employees of Carrongrove Paper Mill, numbering about 200 had their annual excursion to Culross. They were accompanied by their much respected manager, Mr Johnstone.*

QUOTATIONS AND AGREEMENTS ARE BASED UPON SEA-BORNE TRANSIT WHERE POSSIBLE, AND ARE SUBJECT TO THE CONTINGENCIES OF TRANSPORTATION AND STRIKES OR UNAVOIDABLE ACCIDENTS AND DELAYS BEYOND OUR CONTROL.

TO AVOID POSSIBLE DELAYS PLEASE ADDRESS ALL COMMUNICATIONS TO THE COMPANY.

MM LONDON OFFICE: BLACKFRIARS HOUSE, NEW BRIDGE STREET, E.C. 4.

The Carrongrove Paper Coy. Limited.
Proprietors
Esparto Paper Mills Limited

MILL No. 41

DIRECTORS:
SIR FREDERICK BECKER, J.P. (Chairman)
E. B. MONTESOLE.
D. R. HALL CAINE, M.P.
C. SANDEMAN, K.C.
W. M. WALLACE. (Managing Director)

CARRONGROVE PAPER MILLS,
DENNY
STIRLINGSHIRE.

TELEPHONES: SALES DEPT. No. 2 DENNY. PURCHASES DEPT No. 52 DENNY. LONDON OFFICE, No. 2880 CITY.
TELEGRAMS: CARRONGROVE DENNY. CARRONGROVE, LUD, LONDON.
CABLEGRAMS: CARRONGROVE. DENNY. SCOTLAND. CARRONGROVE, LUD, LONDON.
GOODS ADDRESS: CARRONGROVE SIDINGS, L.M.S. RLY. DENNY.

11th August 1924
(Monday)

In reply please quote Ref. W.M.W.

Mr. Robert Moreland,
223 West Scotland Street,
Kinning Park,
GLASGOW.

Dear Sir,

With reference to your interview here with Mr. Herd and myself, we now confirm that we are prepared to give you a start as a Fitter in our Engineering Department all on the lines indicated to you at the interview, - the terms are 1/4d. per hour, and a House is available should you desire same.

You can arrange to start at any time to suit your convenience.

Yours faithfully,
THE CARRONGROVE PAPER COMPANY LTD.,

Managing Director.

NATIONAL SCHEME FOR DISABLED MEN

As the above letter of 11 August 1924 from Managing Director W M Wallace shows, it was fairly standard to offer workers company housing. This housing included electric power generated from the papermaking plant.

Note also the stamp on company stationery of the National Scheme for Disabled Men – papermaking, and more specifically finishing and despatch involved very many repetitive tasks, very capably discharged over many years at Carrongrove by some employees with physical and mental disabilities. Most of these tasks are now done by machine

Courtesy of the National Archives of Scotland GD1/779/13

Staff from Stoneywood paper mill 1888. The photograph appeared in the *World Paper Trade Review*, January 1934

William Wallace, General Manager and Director, Carrongrove 1930

The Wallace era at Carrongrove lasted from 1898 to 1970. William Wallace was appointed manager in 1898 and during his term in office until his death in 1946, he transformed the mill from a two machine loss-making operation to one with four paper machines and four coating machines making a range of products including high class coated art papers. Under his leadership Carrongrove became highly profitable and was the leading mill within the Inveresk Paper Group. His reputation spread well beyond Carrongrove and he became one of the most influential paper makers of the twentieth century. Many paper makers from other leading British mills were trained under Wallace, and engineers from places such as Guardbridge, Dalmore and Henley paper mills would spend hands-on time at Carrongrove. At one stage he was engineering consultant to twenty seven mills throughout the UK.

Carrongrove mill grew to a peak workforce of 740 males females and youths and the management knew almost every worker by name. Expansion of the plant meant it was necessary to build housing and Wallace initiated a house building programme for workers and staff at Fankerton, Stoneywood and Denny.

William Wallace took an interest in the family life of employees and after the General Strike of 1926 when there was no trade union in the mill, he started the Carrongrove Benefit Society to provide pensions and benefits and a system of marriage dowries.

William Wallace's achievements are all the greater for the fact that he was stone deaf – communication had to be written . After his death Wallace's sons, Morgan and Alen took over. They had both been trained in mill management, Morgan under his father and Alen at James Bertram & Co in Edinburgh.

The Carrongrove Benefit Society
was started in 1928.

The company was represented on the Committee by the Manager and senior members of staff and chosen workers made the remainder with a worker as Chairman.

Weekly Contributions
Males 1/– per week
Females 10d per week
Youths 6d per week
The company paid a share equal to the worker's contribution.

Benefits paid
– £1/10/0 per week paid to Males at 65 years
– £1/0/0 per week paid to Females at 60 years
– Marriage dowries were paid to women on a sliding scale according to years' service
– Unemployment Benefit was paid to all workers when laid off owing to shortage of work
– Sick Benefit was paid to all workers off sick

The Society was wound up in 1960 when Government Social Security was increased and workers in receipt of a mill pension were ineligible for state benefits.

29 July 1894
The employees of Carrongrove Paper Co. held their annual excursion to Castle Campbell. The party in 12-horse brakes, made stops at Causewayhead and Alva. Mr & Mrs Johnstone were in the party and a fine outing was had by all.

9 July 1898
Last Saturday, the employees of Messrs John Collins, paper-manufacturers, held their annual excursion to Oban. The party travelled by special train and a pleasant day was spent in boating and exploring the beautiful district.

20 August 1898
The sports in connection with the Stoneywood and Herbertshire Mills took place last Saturday at Stoneywood Park. Competitions were confined to employees and the meeting passed over with fair success. Denny Brass Band was present.

[T Clapperton, *History of Carrongrove Paper Mill* Appendix B]

The *Falkirk Mail* records presentations to mill staff on forthcoming weddings and retirals. On 6 April 1895, Carrongrove employees presented Richard Marshall Junior and his future wife with a handsome marble timepiece and silver jelly cruet. [20] On 13 April 1901, on the occasion of his silver wedding, Hardie Ross,

As well as generating its own power supply, running its own housing, and arranging its own social security, Carrongrove paper mill also ran its own fire service.

The mill fire crew photographed (opposite) in 1940:

Back row (left to right)
Jack McWatt, Bobby Steel, Jimmy Westwater, Jack McKenzie

Front row (left to right)
Jim Dunn, Jackie Muir, Jock Hillhouse, Robert Muir

Boy at back Ivy White

In the front are Driver Ian Martin, Station Officer Frank Simpson and Leading Fireman Walker Herd

cashier at Carrongrove, received a silver ink-stand. [21] On 3 February 1900, Daniel Dewar, plumber at Herbertshire Mill, received a gold-mounted umbrella, Gladstone bag and meerschaum pipe, [22] and on 23 September 1905, Mr Woodbridge, foreman at Stoneywood Paper Mill, received a travelling bag, walking stick and pocket book on his moving to Glasgow to open a fruit and flower business. [23] On 12 September 1978, Alex Couper celebrated 40 years of unbroken service with Inveresk, having joined the Finishing House at Carrongrove in 1938 and then taking up post as Sales Office Manager from 1952. [24]

Other social developments included the establishment of the Carrongrove Mutual Improvement Association in 1894, which met weekly on Saturday nights. Members attended talks on topics as wide ranging as 'The Life of Abraham Lincoln', 'Current Fiction' and 'Co-operation' and took part in debates such as 'Should museums and art galleries be opened on Sundays?' George Johnston both attended and spoke at these functions. [25]

In January 1895, he opened the Carrongrove Penny Savings Bank, a branch of the Glasgow Savings Bank. On the first night more than fifty depositors came forward to deposit over £12 and by the end of the first year, the bank held assets of almost £190. [26]

During this time, the Carrongrove Sick and Benefit Society was also well established. A report of their Annual General Meeting in the *Stirling Journal*, 21 December 1895, declared a balance sheet of £172 6s 10d made up from weekly payments from their 138 membership, and a total sum of £40 17s 6d paid out to the sick and infirm. [27]

John Collins' personal contribution to the welfare of the community was the instigation of a sick benefit scheme, sports facilities for his workers and the establishment of the local Cottage Hospital.

Carrongrove 2000

PAPERMAKING in the twenty first century is still founded on the same basic principles applied in previous centuries. But it has become a highly technical process involving both chemists and engineers whose main concerns, as well as making profits, are to produce the best quality papers as efficiently and economically as possible ensuring the maximum use of resources and raw materials with an eye to energy conservation and environmental issues.

Despite this being the age of the computer and the so-called paperless office, paper is and will remain a fundamental part of our lives whether as books, newspapers and writing materials or as some of the basic necessities of life like toilet tissue, food packaging or stamps.

On 12 October 1990, a £40m management buyout acquired Inveresk Ltd from Georgia Pacific to set up today's company Inveresk plc which is made up of four papermills, Carrongrove, Caldwells, St Cuthberts and Westfield and a European subsidiary, InvereskBV in Germany.

Three years later, the company was floated on the stock exchange. The mill now makes 40,000 tonnes of coated paper per year. The Weir Paper Mill in Alloa was purchased during 1995 thereby increasing the Group's capacity.

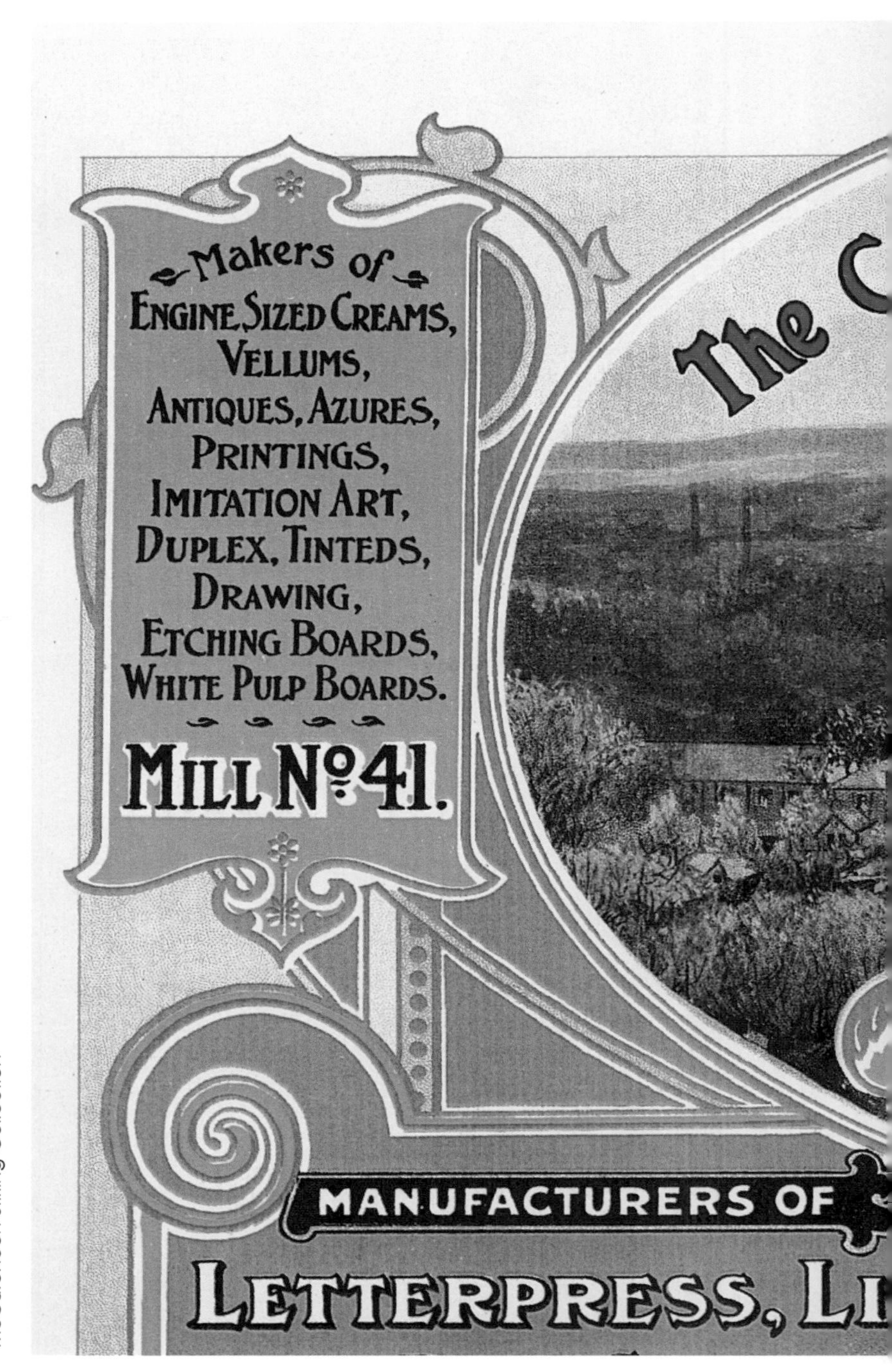

McCutcheon Stirling Collection

Promoting the paper then . . .
an entry in a composite advert placed in
Scotland's Industrial Souvenir (1905) . . .

new
GEMiNi
FROM CARRONGROVE
the graphical board
new Gemini C2-S
new Gemini C1-S

. . . and promoting the paper now . . .
visuals from 1990s trade material
advertising Gemini coated board made at Carrongrove mill

Carrongrove Gemini coated board used for book covers in the home and overseas market, calendars and packaging

C2-S/
ROSE
AGING BOARD
ASSORTED
WEIGHTS
ERING RANGE
GEMINI
Blanc
White
Weiss
Gemini
Printing
C2-S
JANUARY
MARCH
JULY

The stained glass window on the stair way in Carrongrove House, built in the middle of the mill site in 1862 and now used as the site admin offices

Postscript

REVIEWING 200 years of papermaking at Carrongrove and having personally been in the industry for forty years, I can very much relate to the history of papermaking in the mill. Nothing much has changed, or has it?

In 1960 when I joined the industry there were around 94,000 people employed in papermaking in the UK working in some 215 mills around the country. Now we have fewer than 100 mills with only 21,000 people in direct employment and this number continues to reduce. However, the skills of our employees have significantly changed with computer literate technologists now controlling vast machines that can each turn out 250,000 tonnes per year.

David R F Ferguson
General Manager
Carrongrove Mill

Global paper consumption is now around 300 million tonnes per annum and is set to increase by half as much again in the next ten to fifteen years, spurred by technology and by growing literacy and consumer purchasing power in developing economies. Electronic commerce and digital imaging are expanding rapidly but such is the adaptability of paper that it complements rather than competes with the computerised world.

In the UK we consume over 12 million tonnes of paper and board annually with around half of that being produced in this country. Carrongrove Mill, with its one paperboard machine, currently produces a modest 40,000 tonnes per annum. However seventy per cent of this tonnage is exported to some fifteen countries and the last decade has undoubtedly been the most profitable in the recent history of the mill. One wonders what those early entrepreneurs and papermakers would make of these levels of output and to so many export markets.

Over the two hundred years the banks of the Carron have seen many changes in industrial scenery – no mor e so that the quality of the river itself. Those early papermakers would now be amazed at the action taken on effluent treatment to ensure that the river can sustain a healthy fish life. No more does the UK paper industry contribute to a degeneration of the environment with strict controls being applied to all emissions, whether these be to water, land or air. Sixty five per cent of raw material used is waste paper.

We can look back with pride to what our forefathers established and look forward with confidence to Carrongrove Mill continuing as an active contributor to the Scottish economy. Whether this will be for another two hundred years, we can only guess.

David R F Ferguson
General Manager
Carrongrove Mill
May, 2000

Rolls of paperboard come of the end of the production line

(left) the 'wet end' of the twin wire machine at Carrongrove

Notes

(1) *New Statistical Account of Scotland* [Edinburgh, 1845], vol 8, p 115-138.

(2) Alistair G Thomson *The Paper Industry in Scotland 1590-1861* Herbertshire Documents: footnote 32, p 108 (Edinburgh, 1974).

(3) Alistair G Thomson *The Paper Industry in Scotland 1590-1861* Herbertshire Documents: p 100-103 and footnote 32, p 108 (Edinburgh, 1974).

(4) 'Scottish Watermarks', The Quarterly Journal of the British Association of Paper Historians, Issue 19, July 1996 refers to a Carron watermark, (ref: Bower collection BM84101).

(5) The County Voters Roll for the parishes of Denny and Dunipace for 1832, records Gavin Glennie, resident of Milton House, Dunipace, as being in charge of Carron Grove Mill and Robert Weir, paper manufacturer at Herbertshire Mill, resident at Randolphill House.

(6) List of paper mills published by Oliver and Boyd of Edinburgh and Robert Weir of Glasgow.

(7) Falkirk Archives: Forbes of Callendar papers: GD171/1230/14.

(8) Falkirk Archives: Forbes of Callendar papers: GD171/3257.

(9) Tom Clapperton *History of Carrongrove Paper Mill* and researches of J Thomson, Fankerton (Falkirk District Libraries, 1970) p 4.

(10) Ewen Jardine *The History of Paper Mills in Central Scotland*.

(11) NAS: Stirling Sheriff Court records: Register of Deeds: SC67/36/47 ff1-63.

(12) Ewen Jardine *The History of Paper Mills in Central Scotland*

(13) NAS: J Bertram and Son Ltd. Leith: List of papermaking machines constructed, 1847-1972: GD284/25/12.

(14) 'Stout Hearts and Adaptable Skill', from *The Burgh of Denny and Dunipace Official Guide* 1965 p29.

(15) Tom Clapperton *History of Carrongrove Paper Mill* p10.

(16) Tom Clapperton *History of Carrongrove Paper Mill* Appendix A, *Stirling Journal* October 1894.

(17) Tom Clapperton, *History of Carrongrove Paper Mill* Appendix B, *Falkirk Mail* 13 April 1901.

(18) 'Stout Hearts and Adaptable Skill', from *The Burgh of Denny and Dunipace Official Guide* 1965 p 34-35.

(19) Tom Clapperton *History of Carrongrove Paper Mill* Appendix A, *Stirling Journal* 9 June 1894.

(20) Tom Clapperton *History of Carrongrove Paper Mill* Appendix B, *Falkirk Mail* 6 April 1895.

(21) Tom Clapperton *History of Carrongrove Paper Mill* Appendix B, *Falkirk Mail* 13 April 1901.

(22) Tom Clapperton *History of Carrongrove Paper Mill* Appendix B, *Falkirk Mail* 3 February 1900.

(23) Tom Clapperton *History of Carrongrove Paper Mill* Appendix B, *Falkirk Mail* 23 September 1905.

(24) Group Review, The House Journal of Inveresk Group, Winter 1978, p 6.

(25) Tom Clapperton *History of Carrongrove Paper Mill* Appendix A, *Stirling Journal* 17 March 1894.

(26) Tom Clapperton *History of Carrongrove Paper Mill* Appendix A, *Stirling Journal* 29 December 1895.

(27) Tom Clapperton *History of Carrongrove Paper Mill* Appendix A, *Stirling Journal* 21 December 1895.

Bibliography

Clapperton, Tom *History of Carrongrove Paper Mill* and researches of J Thomson, Fankerton (Falkirk District Libraries, 1970)

Hills, Richard L. *Papermaking in Britain, 1488-1988* (London, 1988)

Jardine, Ewen *The History of Paper Mills in Central Scotland*

Thomson, Alistair G. *The Paper Industry in Scotland 1590-1861* (Edinburgh, 1974)

'Scottish Paper Trade History', *The World's Paper Trade Review*, issues 1 March 1912 – 1 May 1914

Waterson, Robert 'Early Papermaking near Edinburgh,' *The Book of the Old Edinburgh Club* vol 25, 1945 p 46-69